About Island Press

Island Press, a nonprofit organization, publishes, markets and distributes the most advanced thinking on the conservation of our natural resources—books about soil, land, water, forests, wildlife and hazardous and toxic wastes. These books are practical tools used by public officials, business and industry leaders, natural resource managers and concerned citizens working to solve both local and global resource problems.

Founded in 1978, Island Press reorganized in 1984 to meet the increasing demand for substantive books on all resource-related issues. Island Press publishes and distributes under its own imprint and offers these services to other nonprofit organizations.

Support for Island Press is provided by Geraldine R. Dodge Foundation, The Energy Foundation, The Charles Engelhard Foundation, The Ford Foundation, Glen Eagles Foundation, The George Gund Foundation, William and Flora Hewlett Foundation, The Joyce Foundation, The John D. and Catherine T. MacArthur Foundation, The Andrew W. Mellon Foundation, The Joyce Mertz-Gilmore Foundation, The New-Land Foundation, The J. N. Pew, Jr. Charitable Trust, Alida Rockefeller, The Rockefeller Brothers Fund, The Rockefeller Foundation, The Florence and John Schumann Foundation, The Tides Foundation and individual donors.

THE ROLLING STONE
ENVIRONMENTAL READER

FOREWORD BY JANN S. WENNER

The

Rolling Stone

Environmental

Reader

ISLAND PRESS

Washington, D.C. · *Covelo, California*

Library of Congress Cataloging-in-Publication Data

The Rolling stone environmental reader / foreword by Jann Wenner.
     p.   cm.
    Includes index.
    ISBN 1-55963-167-8 (cloth : alk. paper). — ISBN 1-55963-166-X (paper : alk. paper)
    1. Ecology. 2. Ecology—Political aspects. 3. Ecology—Public opinion. 4. Human ecology—Public opinion. 5. Pollution.
6. Environmental policy—Public opinion. I. Wenner, Jann.
II. Rolling stone.
QH541.145.R64   1992
333.7—dc20   92-5175
                                        CIP

Rolling Stone environmental reader

CONTENTS

FOREWORD

THE VIEW THAT has characterized most of the environmental stories to appear in *Rolling Stone* in recent years is plain: All of the major institutions of American life are guilty of green crimes. Among those who have argued this most forcefully, and whose articles are included in *The Rolling Stone Environmental Reader*, are William Greider, Howard Kohn and Bill McKibben.

William Greider has been *Rolling Stone*'s national editor for nearly a decade. During these years he has repeatedly confronted the failure of the Washington establishment to meet its responsibilities. This gross negligence of the common welfare, and its meaning for us as a nation, has never been clearer than during the debate over new clean-air legislation. As Greider wrote in the magazine's 1988 year-end issue (the article was headlined "The Dirty Politics of the Environment"): "Nothing illustrates the breakdown of American democracy more starkly than the refusal of the political system to respond . . . to an aroused public's concerns about the environment."

Greider's contempt for those who choose political expediency over civic duty was echoed time and again in the special issue *Rolling Stone* devoted to Earth Day 1990. It was here that Contributing Editor Howard Kohn identified the worst polluter in America: our own federal government. Most *Rolling Stone* readers were already aware of the extent to which our freedoms had been forfeited to satisfy the demands of the national security state. Kohn now revealed how zealous Cold Warriors had also sacrificed the health of countless numbers of our fellow citizens—among them those Americans who have suffered thyroid cancer, leukemia, birth defects and sterility because they had the misfortune to live near one of the government's nuclear-weapons plants.

Like the Cold War, the Reagan years have bequeathed a cruel legacy to the nineties. Yet another *Rolling Stone* contributing editor, Bill McKibben, was among the first observers to draw the connection between the moral climate of the eighties and the continuing destruction of our natural treasures. As McKibben showed most vividly in "Milken, Junk Bonds and Raping Redwoods," the art of the deal in the Gipper's golden age often amounted to another mugging of Mother Nature by harmless-looking businessmen.

There are probably few readers of *Rolling Stone* who would disagree with these writers or most of the others whose stories appear here— Winifred Gallagher, Bill Gifford, Mark Hertsgaard and Tom Hayden among them. Yet lest it be thought that the magazine supports some sort of eco-orthodoxy, also included in *The Rolling Stone Environmental Reader* is P. J. O'Rourke's essay "The Greenhouse Affect." As always, O'Rourke expresses his rock & roll-ribbed Republican views quite forcefully—"a lone voice," as he puts it, "*not* crying in the wilderness, but hollering in the rec room."

It should be clear why these pieces were published in *Rolling Stone.* Like most of the stories included in this book and, I hope, like most of what we run in the magazine, they convey both facts and fervor—solid reporting leavened with passion and not a little disdain for those who consider themselves our leaders. Perhaps the most stirring example here is Tom Horton's "Paradise Lost."

Within days after the *Exxon Valdez* ran aground in Prince William Sound, Horton was in Alaska, assigned by *Rolling Stone* to report on what was clearly one of the biggest stories of the last decade. Horton's job, however, was not to rush back with dramatic footage of oily sea otters, although, tragically enough, he found plenty. Instead, it was his assignment to explore the full significance of this calamity.

When it became all too clear that Horton would need more time in Alaska (and more than the 10,000 words he was originally assigned to write) to fulfill the promise of his early reporting there, he was told to go for it. What Horton brought back was an exhaustive investigation of the makings of a disaster and a compelling account of its aftermath. Yet what was truly revelatory about Tom Horton's story was his uncovering of the spiritual blow the despoiling of the sound dealt the men, women and children who live along its shores.

Had we stopped to think about it for a moment, however, we would have remembered that this too was part of an old American lesson—one that is recalled for us in the piece that closes *The Rolling Stone Environmental Reader*. Paying tribute to Thoreau in this excerpt from *Heaven Is Under Our Feet: Voices for Walden Woods*, Don Henley, E. L. Doctorow and Robert Bly illuminate the deepest meaning of the struggle to save our planet: That we seek to preserve the earth not simply to survive—not simply to drink clean water and breathe clean air—but because this world nourishes the soul of the race.

JANN S. WENNER
Editor and Publisher, Rolling Stone

How We Address Our

PART I

Environmental Problems

The environment holds a strange position in public affairs. The central sources of power have long downplayed it: economists describe it as an "externality," something outside of their calculations; politicians pay lip service and move on to other matters; the public responds favorably to it in opinion polls and then continues to use environmental resources wastefully. But all the while, environmental concerns gain influence as public awareness of their pervasive influence on our lives grows.

But it hasn't always been like this. As Mark Hertsgaard's article attests, the media, even into the late 1980s, did not cover environmental issues, did not even take environmental topics seriously. This attitude, however, is changing. In 1989 the Cable News Network (CNN) took the lead in environmental coverage by appointing Barbara Pyle as a full-time environmental editor, the only one in the country. Other networks have recently followed CNN's example. The print media in particular—in both national and local papers—now devote front-page articles to environmental topics. The media's enhanced coverage testifies to the growing environmental concern among Americans.

Earlier predictions that environmental consciousness would go the way of tie-died shirts and bell-bottoms have yielded to a recognition that the environment has an impact on almost every human activity. Yet even with its momentum gathering, relentless stereotypes still plague the environmental movement. Some stereotypes, like the

pervasive notion of an unavoidable conflict between environmental health and economic growth, have been chipped
away by environmental groups and economists. Other stereotypes, like some environmentalists' tendency to use the
threat of an impending environmental doomsday, may be
deserved.

Environmentalists point to the urgency of the need for
change and support their statements with statistics, science
and emotion. Some industrialists and politicians discount
the warnings and criticize the arguments as being overly
pessimistic. Each side complains of the other's tactics:
naiveté and sloppiness at best, dishonesty and emotional
manipulation at worst. Those left in the middle, the vast
majority of Americans, are torn between their genuine concern for nature and the frustration and guilt spawned by
conflicting information from all sides. P. J. O'Rourke
brings a rare humor to this earnest—some might say too
earnest—debate.

And the debate goes on, with no adequate commitment
from government or industry to environmental quality. Yet
a significant level of energy and attention has been put toward environmental concerns: it is a subject whose time
has finally arrived.

Covering the World; Ignoring the Earth

Mark Hertsgaard

A few years ago one of the *Washington Post*'s senior editors walked into the paper's daily news meeting with what he thought was a pretty big story. Reputable scientists had concluded that there appeared to be a hole in the ozone layer. Widespread use of aerosols, refrigerants and air conditioners had released so many chlorofluorocarbons (CFCs) into the heavens that the stratospheric coating that protects the earth from the full force of the sun's ultraviolet rays had been damaged, perhaps irreparably. An epidemic of skin cancer was but one of the dire foreseeable consequences.

The editor pitched the story for page 1, but *Post* executive editor Ben Bradlee was not impressed. According to one source at the paper, "Bradlee just sat back in his chair, raised his arm and squirted—*psst, psst*—under his arm and laughed. It was inconceivable to Ben that using his Right Guard in the morning could have any environmental effect.

"If you're the national editor," the source continues, "and you know the executive editor is going to raise his arm and act like he's spraying underarm deodorant when you pitch an ozone story, you're not going to give it a real hard push."

Bradlee says he now tells this story on himself and adds: "I did say once that if you sprayed your underarms in the basement of a New York apartment house, I had trouble seeing how that could hurt the ozone layer. I'm not sure I understand it to this day."

Though clearly one of the great newspaper editors of his time, Ben Bradlee was, on the issue of the environment, typical of his profession. "Even five years ago [1984] the environment was not in the forefront of our minds at any of the big papers," says the *Post*'s national editor, Karen DeYoung. "Certainly it is now. I think people started getting really scared. When people began to realize that this was real stuff and concerned us now—not ten generations down the road—it made a difference."

Clearly, the summer of 1988 deserves much of the credit for this shift in journalistic, and public, consciousness. It was a hellish season. If you were a farmer, you watched your land bake hard as stone. If you lived in the city, you boiled like a turnip in a pot. But the summer did have one salutary result: It woke people up to the dangers of the greenhouse effect, to the probability that the earth is gradually overheating from all the smoke and soot Industrial Man has spewed into the atmosphere. Scientists had long feared such a prospect. They theorized that global warming might bring widespread drought and famine, unleashing Genesis-like floods the world over, as ice caps melted and oceans submerged their shores. But because few people in positions of influence paid much attention, the public was not alerted.

Not until a top NASA scientist, James Hansen, testified before

Congress in June of last year [1989] that the greenhouse effect was no longer merely theory but fact, did the media really take notice. The *New York Times* put the story above the fold on page 1 and gave the subject extensive play the rest of the summer. So, to lesser degrees, did other news organizations. Finally, the greenhouse effect had made it onto the national agenda.

Except for nuclear war—the ultimate environmental threat—global warming is probably the single most worrisome storm cloud on our bleak environmental horizon. Yet for years no one but environmentalists seemed to care. "When the greenhouse effect began making headlines last summer [1988], someone showed me the first article I ever wrote on the subject," recalls Philip Shabecoff, author of the *Times'* front-page greenhouse story. "It was from 1979, and it said many of the same things about what was likely to happen as last year's stories did. It ran on page 42. I guess you could say it took nine years for the greenhouse story to graduate to page 1."

Technically speaking, the media's sudden interest in the greenhouse effect rested on shaky science. The heat of 1988 could not be blamed directly on the greenhouse effect; weather is the result of too many variables to draw any single cause-and-effect conclusion. Hansen, who works at the Goddard Institute for Space Studies, was careful to make this point; his worry was the weather in the year 2020, not one hot summer in 1988. Nevertheless, had the weather a year and a half ago [summer 1988] not been so ghastly, it's doubtful journalists, politicians or the public would have listened any more closely to Hansen than they had to previous warnings. After all, Hansen himself had given much the same congressional testimony two years earlier with no apparent effect. Major reports by the National Academy of Sciences, the Environmental Protection Agency and the Massachusetts Institute of Technology dating back as far as 1971 had likewise failed to attract sustained media coverage or engage the interest of more than a few lawmakers.

Of course, 1988 was also the year when forest fires ravaged Yellowstone, the Mississippi dried up, garbage and medical wastes soiled East Coast beaches, and pollution-weakened seals died en masse in the North

Sea. Along with such recurrent dangers as acid rain, toxic wastes and ozone depletion, these disasters further concentrated the collective journalistic mind on the gathering environmental crisis.

As a result, the environment moved onto the front page last year as never before. The *New York Times* sounded the alarm about the health hazards posed by the breakdown of federal nuclear-weapons production plants; *60 Minutes* sparked a national uproar with its report on the cancer risks for children eating apples sprayed with Alar. Most striking of all, *Time*, in a rare departure, chose Endangered Earth as its Man of the Year, devoting nearly an entire issue to the plight of the planet.

Certainly, over the last eighteen months, the press has begun doing a better job of covering the environment. But as heartening as the media's new-found interest is, it is not enough. If our children, to say nothing of *their* children, are to inhabit a survivable world, some very fundamental changes must be made. The success of this transformation will depend in no small measure on the news media. Because of their enormous power to shape public opinion, the men and women who are in charge of the media have an immense responsibility in the struggle to avert ecological catastrophe.

"We don't have time for the traditional approach to education—training new generations of teachers to train new generations of students—because we don't have generations, we have years," says Lester Brown of the Worldwatch Institute. "The communications industry is the only instrument that has the capacity to educate on the scale needed and in the time available."

Why didn't the press trumpet the urgency of the greenhouse problem earlier?

The simple fact is that the environment has traditionally not been seen as a very important news story by reporters or by their editorial and executive superiors. Even now, at most of the nation's major news organizations, the environment is not considered a bona fide beat. Neither CBS nor NBC news has even one full-time environmental correspondent; ABC News only named its first two earlier this year. All this stands in marked contrast to CNN, where owner Ted Turner's strong personal

6

interest has ensured extensive and continuing coverage. Turner's eight-member environmental unit, established in 1980, is expected to produce a story a day for the network, according to environmental editor Barbara Pyle.

CNN notwithstanding, when it comes to getting their work on the air or in the paper, environmental reporters often find themselves victims of a journalistic value system that underplays ecological issues. "Particularly in Washington the big beats are the White House, the State Department, the Pentagon and the related issues on Capitol Hill," says David Goeller, who covered environmental issues for the Associated Press for four years before joining Environmental Action, an advocacy group, in 1988. "I think the greenhouse effect and CFCs are a hell of a lot more important to the future of the world than the U.S. budget deficit. I don't want to say the *Times* and *Post* aren't doing a good job, though I'd say the *Times* is doing a better job. But at both papers the environmental reporters are competing for space against whatever this week's diplomatic crisis is."

The *Times* and the *Post* are the two most politically powerful newspapers in the country, not simply because they are read by the movers and shakers in Washington, but because their coverage tends to shape the news agenda of the networks and the rest of the press as well. Stories the *Times* and the *Post* downplay have little chance of reaching a national audience. Environmental stories in particular have often suffered this fate, largely because both papers were until recently run by editors who were not exactly ardent ecologists.

Nevertheless, even those distressed by Ben Bradlee's previous attitude say he has "turned around a lot in the last year" following a trip to Brazil with a group of U.S. senators. National editor DeYoung stresses that the *Post* recently beefed up its environmental staff to three people and, for the first time, hired a full-time reporter on energy policy. "There's much greater receptivity now to putting those stories on the front page," she says.

Despite these reforms, environmentalists consistently rate the *Times* as superior to the *Post*, singling out the *Times'* Shabecoff as the main reason. Shabecoff modestly explains that he enjoys one major advantage over his competitors: He's been working this beat for more than ten years now. "These are complex issues," he says. "But the turnover of environmental reporters at most news organizations is very rapid. Except for Mary Hager at *Newsweek*, there's nobody that's been on the beat more than three or four years. You really can't do it that way. You have to build up a storehouse of knowledge."

Shabecoff's career at the *Times* has not been without frustrations, however. One of them was Abe Rosenthal, the editor who ran the paper for seventeen years prior to his retirement in 1986. Shabecoff, while declining to comment on Rosenthal directly, acknowledges that during Rosenthal's tenure the environment "quite obviously was not treated as an issue of equal import to national security or the economy, and it did take a quantum leap in coverage after Max [Frankel] became editor. . . . I remember having written a long article about the role of the World Bank and other financial institutions in contributing to environmental degradation in third-world countries through their lending practices. That story languished unpublished for a year, because it didn't fit any of the normal news categories. The week Max took over, it was at the top of page 1."

It was under Frankel in 1988 that the *Times* published its long-running exposé of the nuclear-weapons industry. The investigative pieces revealed that production facilities run by Du Pont and other corporate contractors for the Department of Energy had leaked ruinous amounts of radioactive waste into surrounding land and water supplies. Certainly, no one could complain that this story was underplayed. Keith Schneider, the principal reporter on the investigation, notes that in the three months following the initial four-part, front-page series, he and five other *Times* reporters "wrote over 100 stories on this, and we still are, and it's still getting on our front page." Schneider does admit, however, that what accounted for the story's big play was not so much the health and environmental hazards as "the fact that this had a national-security angle."

The troubling question is, why did it take the national press so long to recognize this story in the first place? As Peter Dykstra of the environmental organization Greenpeace observes: "Nobody in the press paid any attention to the weapons plants until the *Times* made something of a crusade about it. I'm glad they did, but why couldn't that story stand on its own? All those problems existed ten years ago, but nobody knew about it except about 200,000 environmentalists, who couldn't get anyone in the press to listen to them."

Now that the environment is a hot topic, the favorite new buzzword within the press seems to be *green*. It's a word that reporters and headline writers have appropriated from the ecologically minded political parties that have sprung up in Europe and elsewhere over the past decade. As it happens, the green parties espouse some pretty radical political ideas. Not only do they unconditionally oppose nuclear power plants and armaments, but they also challenge the entire ideology of modern industrial society. They strongly reject the conventional equation of economic growth with prosperity and regard *technological progress* as an oxymoron. They advocate a complete reordering of the economy in deference to the environment. In short, although they often work within the system, they are revolutionaries.

Yet now that it's smart politics to do so, any politician who pays lip service to environmental concerns gets labeled a green by the press. *Newsweek*, the *Los Angeles Times* and many other news organizations heralded the meeting of the world's seven biggest industrial powers in Paris this July [1989] as "the Green Summit," despite the fact that the assembled heads of government stood for everything real greens oppose and produced an environmental agreement woefully short on concrete action.

The *New York Times Magazine* used the same gimmick a few weeks later in a cover story on the EPA administrator, William Reilly, titled "Greening the White House." Proclaiming that "the courtly new environmental chief speaks softly—but he obviously has the president's ear," the *Times* piece was but one of many mainstream news stories in recent months to accept the absurd notion that George Bush is an environmentalist. This is the same Bush who, as vice president, served in the most environmentally hostile administration of the modern era. In fact, Bush was a key player in Ronald Reagan's egregious assaults on the environment.

As chairman of the President's Task Force on Regulatory Relief, Bush oversaw the repeal or weakening of federal standards governing auto pollution, energy conservation and clean water, among others. His chief of staff, Craig Fuller, was an inside player in the celebrated conflict-of-interest scandal in the early eighties involving the EPA Superfund program for hazardous-waste disposal. Now, as president, Bush remains sufficiently loyal to his corporate past and allies to push for congressional passage of a badly flawed Clean Air Act [passed in 1990]. Not only does it encourage companies to buy and sell the right to pollute, it also moves much too slowly on improving automotive fuel efficiency.

Despite all this, the White House has managed to portray George Bush as an environmentally sensitive president. "It's called the law of relativity," says Ralph Nader. "He's being compared with Reagan, and all he can do is go up. Show the press a trend, and they'll write it." And not just the press. The Washington environmental lobbyists and policy experts to whom reporters turn for balancing remarks to the administration line have given Bush, in Nader's words, "a long honeymoon." Norman Dean of the National Wildlife Federation agrees that "a lot of the positive press has to do with a reflection of how environmental groups view Bush. We're trying to be politically realistic." In other words, environmentalists are unwilling—so far—to offend Bush with too much criticism.

Other environmentalists complain that when they do dispute presidential policy, their comments get buried deep inside news stories and sometimes shut out altogether. "The typical White House story will quote the administration at length and then have one paragraph at the bottom saying environmental groups were unhappy, the gist being that it's important to know what Bush thinks and less important to know what the environmental movement thinks," says Diane MacEachern, president of a Washington public-relations firm that has represented several environmental groups.

Giving the president top—and often sole—billing is a time-honored if misguided convention of mainstream journalism. A second convention that distorts environmental coverage is the media's ahistorical, crisis-driven approach. News is defined as what's happening today, not what might happen tomorrow. As a result, the media's favorite environmental stories are disaster stories, preferably ones boasting gripping TV visuals: The Exxon Alaska oil spill, with its oozing shorelines and sad-faced sea otters, is but the most recent case in point.

Of course, such events deserve comprehensive coverage. But the overriding fixation on the immediate and the spectacular compromises responsible coverage of the environment. Pegging news coverage only to events means that reporters don't arrive on the scene until after the damage is done: The problem at hand is therefore never exposed until it is too late to do much but (try to) clean up the mess. Moreover, some of the most ominous hazards we face—such as global warming and ozone depletion—have been decades in the making and will be years in manifesting themselves fully. That they are gradual rather than sudden processes makes them no less deadly. They just don't yield daily photo opportunities tailored to the eye-blinking attention span of the media. How do you take a picture of the earth getting hotter?

"There isn't a 'Stop the presses!' kind of development on the environment story every day," declares Tom Winship, former editor of the *Boston Globe.* "This is not event coverage. We need to persuade the media to cover the environmental story consistently. Sure, it's a slow story, but they've got to change their attitudes about what makes a story."

Winship's point is well taken. In one respect, though, I think the media would do well to adopt even more of a crisis mentality. Just as Walter Cronkite used to make a point of reminding viewers at the end of every broadcast of *The CBS Evening News* about the American citizens being held hostage in Iran—"And that's the way it is, the twenty-third day of American hostages in captivity," the anchorman would intone— so today's journalists could make a point of reminding people about the environmental crisis that in effect holds each one of us hostage.

Rather than bore us with the ups and downs of the stock market every night, why couldn't newscasts occasionally lead into commercials with graphics depicting that day's smog levels in major world cities, or number of acres of rain forest destroyed, or tons of topsoil eroded, or plant and animal species wiped out?

"We know what changes there were yesterday in the spot market price of oil, but the fact that many tons of carbon were released into the atmosphere by burning fossil fuels, which is the really important energy data, does not get mentioned," says Lester Brown. "This represents a dangerously outdated set of values about the use of public resources."

Of course, news organizations could make such changes easily enough. The obstacles are not technical but ideological. To many in the American press, such journalism smacks of advocacy, partisanship, editorializing—all the sins an objective professional is supposed to resist. Yet even within the parameters of mainstream journalism, argues Tom Winship, "there are legitimate ways to do advocacy journalism— how prominently you display a given story, how often you cover it, how much editorial support you offer." Winship points to *Time*'s Planet of the Year issue as a model. "Nothing had a greater impact on establishment journalism's treatment of the environment than that piece. It was useful, crusading journalism such as we haven't seen in years."

What was remarkable about the Planet of the Year issue was not only the apocalyptic fervor with which *Time* described the earth's problems but also its enumeration of specific cures. Each section of the report included a box headlined "What Nations Should Do," which listed four or five concrete steps to counter global warming, the extinction of species and other environmental threats. At the end of the report was a full page titled "What the U.S. Should Do." Some proposals were insufficiently bold, such as the call for automobile fuel efficiencies of forty-five miles per gallon by the year 2000 (rather than, say, phasing out gasoline altogether); others were wrongheaded, such as the dubious notion of developing safer designs for nuclear reactors. But most were steps in the right direction. Besides, one need not endorse all of *Time*'s suggestions to appreciate the value of journalism that does not merely describe problems but also prescribes solutions. Call it advocacy journalism, but at this late date we need as much of it as possible.

Nader argues that the press should do more reporting of environmental success stories. "There's lots happening around the country with solar energy they could be reporting," says Nader, "but it's not a graphic story, and there's no official support for solar from the Department of Energy. Plus, it's easier to cover conflict at the end source, when pollution is

exposing itself to human beings, than to cover the emerging displace-
ment technologies. . . . The press is not up to reporting horizons."

It's true the press generally shuns playing a leadership role, though
that's not the only reason aggressive coverage of environmental alternatives is rare. Another impediment is that meaningful solutions to these
problems will require challenging some of America's most powerful
institutions and interest groups.

"What's needed now is an analysis of what it's going to take to act,"
says environmentalist and writer Barry Commoner. "If you go back to its
origin, the environmental problem originates in the means of production. We build the wrong kind of cars, so we have smog. We use
chemicals to raise our food, so we get water pollution. . . . In our
capitalist system the owner of the means of production is free to produce
whatever he wants, however he wants. So the quest to improve the
environment immediately raises a very fundamental—in fact, taboo—
question: Does society have the right to intervene in the rights of
property owners? Until the necessity for this action is made clear, the will
to do so will not develop."

But as central pillars of the American establishment themselves, the
nation's leading news organizations are decidedly unreceptive to ideas
challenging capitalist orthodoxy. Nor do they tend to do much tough
coverage of their fellow corporate giants. As Nader observes, "Look at all
the stories on the destruction of the Amazon rain forest. Do you ever see
the names of any multinational corporations mentioned?"

The question facing us may well be, as Daniel Zwerdling, the former
environmental reporter for NPR phrases it, "Can we survive with the
automobile?" But what newspaper or network will press this issue and
risk alienating companies responsible for tens of millions of dollars'
worth of advertising? Left to their own devices, journalists will instead
content themselves with simply relaying the debate on Capitol Hill,
where all sides assume the automobile is here to stay and argue only over
how efficient its engine must be. (One brilliant, if fatalistic, exception to
this rule was a recent [1989] column by Russell Baker arguing that
America can't survive its auto dependence, but it's too late to do much
about it.)

13

It may be utopian to urge the news media to lead the charge on the environmental issue. As *New York Times* columnist Tom Wicker, one of the few mainstream commentators to champion ecological concerns, says, "It's not in the nature of institutions like the *Times* or *Post* to stir up the community on very broad social questions."

But saving the environment is hardly a controversial cause. Indeed, the public may well be ahead of the so-called opinion leaders in government and the press on this issue, just as it was a few years ago when there was clear support for a bilateral nuclear-weapons freeze. In any case, the press has an obligation to illuminate the dimensions and roots of the environmental crisis and identify and analyze potential solutions. In keeping with this spirit, here are some specific steps that journalists and news organizations could take to improve coverage:

1. Make the environment a priority. Now that the cold war is over, the struggle to save the planet figures to be the biggest news story of the next twenty years. The environment clearly deserves as much attention as the drug issue, another story that offers no daily news pegs but still gets saturation coverage.

2. Make the environment a high-prestige beat, not a training ground or exile post, and make it worth a reporter's while to stay on it. Educate reporters on all beats, be they national security, finance or local politics, to recognize the environmental aspects of their stories.

3. Take the initiative. Good environmental reporting is often investigative; give journalists time and freedom from daily stories to pursue ambitious topics.

4. Analyze the political economy of environmental change and expose obstacles to reform. One story idea: Research the campaign contributions of those members of Congress serving on the committees considering the Clean Air Act to see how much money from the oil, auto and electric-utility industries they've received.

5. Cover America's dissidents, too. Broaden the range of "quotable" sources to include such experts as Barry Commoner. And pay attention to the environmental movement, especially the grass-roots groups on the front lines of the struggle.

6. Remember above all else how little time there is. At this late date, to cover the world while ignoring the earth is sheer folly.

The Greenhouse Affect

P. J. O'Rourke

*I*f the great outdoors is so swell, how come the homeless aren't more fond of it?

There. I wanted to be the one person to say a discouraging word about Earth Day—a lone voice *not* crying in the wilderness, thank you, but hollering in the rec room.

On April 22 [1990]—while everybody else was engaged in a great, smarmy fit of agreeing with himself about chlorofluorocarbons, while *tout le* rapidly-losing-plant-and-animal-species *monde* traded hugs of unanimity over plastic-milk-bottle recycling, while all of you praised one another to the ozone-depleted skies for your brave opposition to coastal flooding and every man Jack and woman Jill told child Jason how bad it is to put crude oil on baby seals—I was home in front of the VCR snacking high on the food chain.

But can any decent, caring resident of this planet possibly disagree with the goals and aspirations embodied in the celebration of Earth Day? No.

That's what bothers me. Mass movements are always a worry. There's a whiff of the lynch mob or the lemming migration about any overlarge gathering of like-thinking individuals, no matter how virtuous their cause. Even a band of angels can turn ugly and start looting if enough angels are hanging around unemployed and convinced that succubi own all the liquor stores in heaven.

Whenever I'm in the middle of conformity, surrounded by oneness of mind, with people oozing concurrence on every side, I get scared. And when I find myself agreeing with everybody, I get really scared.

Sometimes it's worse when everybody's right than when everybody's wrong. Everybody in fifteenth-century Spain was wrong about where China is, and as a result, Columbus discovered Caribbean vacations. On the other hand, everybody in fifteenth-century Spain was right about heresies: They're heretical. But that didn't make the Spanish Inquisition more fun for the people who were burned at the stake.

A mass movement that's correct is especially dangerous when it's right about a problem that needs fixing. Then all those masses in the mass movement have to be called to action, and that call to action better be exciting, or the masses will lose interest and wander off to play arcade games. What's exciting? Monitoring the release into the atmosphere of glycol ethers used in the manufacture of brake-fluid anti-icing additives? No. But what about some violence, an enemy, someone to hate?

Mass movements need what Eric Hoffer—in *The True Believer*, his book about the kind of creepy misfits who join mass movements—calls a "unifying agent."

"Hatred is the most accessible and comprehensive of all unifying agents," writes Hoffer. "Mass movements can rise and spread without belief in a God, but never without belief in a devil." Hoffer goes on to cite historian F. A. Voigt's account of a Japanese mission sent to Berlin in 1932 to study the National Socialist movement. Voigt asked a member of the mission what he thought. He replied, "It is magnificent. I wish we could have something like it in Japan, only we can't, because we haven't got any Jews."

The environmental movement has, I'm afraid, discovered a unifying agent. I almost said "scapegoat," but scapegoats are probably an endangered species. Besides, all animals are innocent, noble, upright, honest and fair in their dealings and have a great sense of humor. Anyway, the environmental movement has found its necessary enemy in the form of that ubiquitous evil—already so familiar to Hollywood scriptwriters, pulp-paperback authors, minority spokespersons, feminists, members of ACT UP, the Christic Institute and Democratic candidates for president: Big Business.

Now, you might think Big Business would be hard to define in this day of leveraged finances and interlocking technologies. Not so. Big Business is every kind of business except the kind from which the person who's complaining draws his pay. Thus the rock-around-the-rain-forest crowd imagines record companies are a cottage industry. The Sheen family considers movie conglomerates to be a part of the arts and crafts movement. And Ralph Nader thinks the wholesale lobbying of Congress by huge tax-exempt, public-interest advocacy groups is akin to working the family farm.

This is why it's rarely an identifiable person (and, of course, never you or me) who pollutes. It's a vague, sinister, faceless thing called industry. The National Wildlife Federation's booklet on toxic-chemical releases says, "Industry dumped more than 2.3 billion pounds of toxic chemicals into or onto the land." What will "industry" do next? Visit us with a plague of boils? Make off with our firstborn? Or maybe it will wreck the Barcalounger. "Once-durable products like furniture are made to fall apart quickly, requiring more frequent replacement," claims the press kit of Inform, a New York–based environmental group that seems to be missing a few sunflower seeds from its trail mix. But even a respectable old establishmentarian organization like the Sierra Club is not above giving a villainous and conspiratorial cast to those who disagree with its legislative agenda. "For the past eight years, this country's major polluters and their friends in the Reagan administration and Congress have impeded the progress of bills introduced by congressional Clean Air advocates," says the Sierra Club's 1989–90 conservation campaign press package. And here at *Rolling Stone*—where we are so opposed to the profit motive that we work for free, refuse to accept advertising and give the magazine away at newsstands—writer Trip Gabriel, in his *Rolling Stone* 571 article "Coming Back to Earth: A Look at Earth Day 1990," avers, "The yuppie belief in the sanctity of material possessions, no matter what the cost in resource depletion, squared perfectly with the philosophy of the Reaganites—to exploit the nation's natural resources for the sake of business."

Sure, "business" and "industry" and "their friends in the Reagan administration and Congress" make swell targets. Nobody squirts sulfur dioxide into the air as a hobby or tosses PCBs [polychlorinated biphenyls] into rivers as an act of charity. Pollution occurs in the course of human enterprise. It is a by-product of people making things like a living, including yours. If we desire, for ourselves and our progeny, a world that's not too stinky and carcinogenic, we're going to need the technical expertise, entrepreneurial vigor and marketing genius of every business and industry. And if you think pollution is the fault only of Reaganite yuppies wallowing in capitalist greed, then go take a deep breath in Smolensk or a long drink from the river Volga.

Sorry, but business and industry—trade and manufacturing—are inherent to civilization. Every human society, no matter how wholesomely primitive, practices as much trade and manufacturing as it can figure out. It is the fruits of trade and manufacturing that raise us from the wearying muck of subsistence and give us the health, wealth, education, leisure and warm, dry rooms with Xerox machines—all of which allow us to be the ecology-conscious, selfless, splendid individuals we are.

Our ancestors were too busy wresting a living from nature to go on any nature hikes. The first European ever known to have climbed a mountain for the view was the poet Petrarch. That wasn't until the fourteenth century. And when Petrarch got to the top of Mont Ventoux, he opened a copy of Saint Augustine's *Confessions* and was shamed by the passage about men "who go to admire the high mountains and the immensity of the oceans and the course of the heaven . . . and neglect themselves." Worship of nature may be ancient, but seeing nature as cuddlesome, hug-a-bear and too cute for words is strictly a modern fashion.

The Luddite side of the environmental movement would have us destroy or eschew technology—throw down the ladder by which we climbed. Well, nuts (and berries and fiber) to you, you shrub huggers. It's time we in the industrialized nations admitted what safe, comfortable and fun-filled lives we lead. If we don't, we will cause irreparable harm to the disadvantaged peoples of the world. They're going to laugh themselves to death listening to us whine.

Contempt for material progress is not only funny but unfair. The average Juan, Chang or Mobutu out there in the parts of the world where every day is Earth Day—or Dirt and Squalor Day anyhow—would like to have a color television too. He'd also like some comfy Reeboks, a Nintendo Power Glove and a Jeep Cherokee. And he means to get them. I wouldn't care to be the skinny health-food nut waving a copy of *50 Simple Things You Can Do to Save the Earth* who tries to stand in his way.

There was something else keeping me indoors on April 22 [1990]. Certain eco-doomsters are not only unreasonable in their attitude toward business, they're unreasonable in their attitude toward reason. I can understand harboring mistrust of technology. I myself wouldn't be inclined to wash my dog in toluene or picnic in the nude at Bhopal. But to deny the validity of the scientific method is to resign your position as a sentient being. You'd better go look for work as a lungwort plant or an Eastern European Communist-party chairman.

For example, here we have the environmental movement screeching like New Kids on the Block fans because President Bush asked for a bit more scientific research on global warming before we cork everybody's Honda, ban the use of underarm deodorants and replace all the coal fuel in our electrical-generating plants with windmills. The greenhouse effect is a complex hypothesis. You can hate George Bush as much as you like and the thing won't get simpler. "The most dire predictions about global warming are being toned down by many experts," said a *Washington Post* story last January [1990]. And that same month the *New York Times* told me a new ice age was only a couple of thousand years away.

On the original Earth Day, in 1970—when the world was going to end from overcrowding instead of overheating—the best-selling author of *The Population Bomb*, Dr. Paul Ehrlich, was making dire predictions as fast as his earnestly frowning mouth could move. Dr. Ehrlich predicted that America would have water rationing by 1974 and food rationing by 1980; that hepatitis and dysentery rates in the United States would increase by 500 percent due to population density; and that the oceans could be as dead as Lake Erie by 1979. Today Lake Erie is doing better than Perrier, and Dr. Ehrlich is still pounding sand down a rat hole.

Now, don't get me wrong: Even registered Republicans believe ecological problems are real. Real solutions, however, will not be found through pop hysteria or the merchandising of panic. Genuine hard-got knowledge is required. The collegiate idealists who stuff the ranks of the environmental movement seem willing to do absolutely anything to save the biosphere except take science courses and learn something about it. In 1971, American universities awarded 4,390 doctorates in the physical sciences. After fifteen years of youthful fretting over the planet's future, the number was 3,551.

It wouldn't even be all that expensive to make the world clean and prosperous. According to the September 1989 issue of *Scientific American*, which was devoted to scholarly articles about ecological issues, the cost of achieving sustainable and environmentally healthy worldwide economic development by the year 2000 would be about $729 billion. That's roughly fourteen dollars per person per year for ten years. To translate that into sandal-and-candle terms, $729 billion is less than three-quarters of what the world spends annually on armaments.

The Earth can be saved, but not by legislative fiat. Expecting President Bush to cure global warming by sending a bill to Congress is to subscribe to that eternal fantasy of totalitarians and Democrats from Massachusetts: a law against bad weather.

Sometimes I wonder if the fans of eco-Armageddon even want the world's problems to get better. Improved methods of toxic-chemical incineration, stack scrubbers for fossil fuel power plants, and sensible solid-waste management schemes lack melodramatic appeal. There's nothing apocalyptic about gasohol. And it's hard to picture a Byronic hero sorting his beer bottles by color at the recycling center. The beliefs of some environmentalists seem to have little to do with the welfare of the globe or of its inhabitants and a lot to do with the parlor primitivism of the Romantic Movement.

There is this horrible idea, beginning with Jean Jacques Rousseau and still going strong in college classrooms, that natural man is naturally good. All we have to do is strip away the neuroses, repressions and Dial soap of modern society, and mankind will return to an Edenic state. Anybody who's ever met a toddler knows this is soy-protein baloney. Neolithic man was not a guy who always left his campsite cleaner than he found it. Ancient humans trashed half the map with indiscriminate use of fire for slash-and-burn agriculture and hunting drives. They caused desertification through overgrazing and firewood cutting in North Africa, the Middle East and China. And they were responsible for the extinction of mammoths, mastodons, cave bears, giant sloths, New World camels and horses and thousands of other species. Their record on women's issues and minority rights wasn't so hot either. You can return to nature, go back to leading the simple, fulfilling life of the hunter-gatherer if you want, but don't let me catch you poking around in my garbage cans for food.

Then there are the beasts-are-our-buddies types. I've got a brochure from the International Fund for Animal Welfare containing a section called "Highlights of IFAW's History," and I quote: "1978—Campaign to save iguanas from cruelty in Nicaraguan marketplaces—people sew animals' mouths shut."

1978 was the middle of the Nicaraguan civil war. This means that while the evil dirt sack Somoza was shooting it out with the idiot Marxist Sandinistas, the International Fund for Animal Welfare was flying somebody to besieged Managua to check on lizard lips.

The neo-hippie-dips, the sentimentality-crazed iguana anthropo-

morphizers, the Chicken Littles, the three-bong-hit William Blakes—
thank God these people don't actually go outdoors much, or the environment would be even worse than it is already.

But ecology's fools don't upset me. It's the wise guys I'm leery of.
Tyranny is implicit in the environmental movement. Although Earth
Day participants are going to be surprised to hear themselves accused of
fascist tendencies, dictatorship is the unspoken agenda of every morality-
based political campaign. Check out Moslem fundamentalists or the
right-to-lifers. Like abortion opponents and Iranian imams, the environmentalists have the right to tell the rest of us what to do because they are
morally correct and we are not. Plus the tree squeezers care more, which
makes them an elite—an aristocracy of mushiness. They know what's
good for us even when we're too lazy or shortsighted to snip plastic six-
pack collars so sea turtles won't strangle.

The Dirty Politics
of the Environment

The term competitiveness has long been used by business and politicians to undermine environmental regulations. Recently, environmentalists have taken up the capitalist battle cry as they come to realize that the language of economics—bottom line, growth, jobs—holds sway in Congress.

Environmentalists now argue that strict environmental standards for American corporations can improve their competitiveness in international markets. In many cases, higher environmental standards have worked to make companies more competitive. Unfortunately, many other industries lobby Congress to loosen or stem the flow of environmental regulation.

But ironically, as William Greider points out, the biggest impediment to strict environmental legislation may be the nature of politics itself. Greider witnessed how "home-state economic issues" and "provincial political feuds" have shaped, and will continue to shape, the quality of environmental legislation passed in this country. He covered the creation of the most momentous environmental legislative effort of the eighties, the Clean Air Act. (This was enacted into law in 1990 in much the shape that Greider's second article predicts.) Perhaps, however, the most bitter irony of all is revealed by Howard Kohn, who reports that the federal government itself is the biggest polluter in the country. These articles explore the forces that are responsible for the quality of America's environmental legislation.

The inevitable tensions between economic and environmental concerns will likely continue to be played out in a highly politicized arena. Like Vice-President Bush before him, Dan Quayle heads a White House council that fights regulatory actions put forward by Congress, the EPA and other agencies. Quayle's Council on Competitiveness has targeted environmental standards more than any other regulatory area. The Council bases its actions on the belief that whatever is good for the environment is overwhelmingly bad for business.

The Council is not alone. The Bush administration continues to back the National Energy Strategy (NES) put forward by the Department of Energy. Only in February 1992, after the EPA's announcement that the ozone layer over North America is depleting at an unexpectedly rapid rate, did the Bush administration reverse its long-standing opposition to emissions cuts.

Some environmentalists hope that even if domestic political pressures fight environmental progress, international pressure may come to bear more strongly on U.S. regulatory procedures. The NES is drastically out of step with worldwide trends in energy and development. European countries are taking pro-environment stances, and European unification has improved environmental standards rather than lowered them.

Overall, new pressures have joined traditional ones to push for a healthy environment. Environmental groups now show the ability to express themselves in the economic terms that carry so much weight with the government. Yet familiar barriers remain, and it will be an unexpected day indeed when the political atmosphere moves away from its consistent posture of environmental confusion.

The Dirty Politics of the Environment

William Greider

*E*veryone talked about the weather in 1988, but nobody in Washington did anything about it. All over the country this year, hot weather, droughts and smog alerts in major cities were consuming topics of conversation—and also sources of great anxiety. Perhaps the most ominous thing about all the bad environmental news was that it seemed to confirm what scientific authorities had been predicting for a long time—that staggering amounts of man-made pollution were causing the formation of a dangerous global hothouse, the infamous greenhouse effect.

North Sea seals killed by ocean pollution; medical garbage washed up on Jersey beaches; a North Carolina red-spruce forest killed by air pollution and acid rain. These and many other unsettling events scared the hell out of average citizens. But Mother Nature's message did not get through to the politicians in Washington. The nation's capital suffered, too, with smog and record-breaking high temperatures, and several measures were passed to deal with specific problems. But on the whole, it was politics as usual for Congress and the White House, for Democrats and Republicans.

The headline for this report on the politics of the environment might read "Washington Diddles while World Burns." Only the story really isn't terribly funny. Nothing illustrates the breakdown of American democracy more starkly than the refusal of the political system to respond this year to an aroused public's concerns about the environment. For months now citizens around the country have been brooding about the hole in the ozone layer over Antarctica, the acid rain killing Eastern forests and lakes, the "brown cloud" over Denver and the terrifying greenhouse effect developing in the polluted atmosphere. But Washington took no notice.

For the eighth year in a row, Congress failed to pass new clean-air legislation. In the Senate a bill that was intended to attack the acid-rain problem was reported from committee in November 1987. But Democratic leaders held it back from a floor vote while they dickered privately among themselves, trying to protect their home-state economic interests. The bill died on the clerk's desk. In the House of Representatives, they didn't even get that far: lots of back-room bargaining but no bill, no embarrassing roll calls. In fairness to the Democrats, environmental bills are extremely difficult for Congress to pass on its own, when the man in the White House has no interest in such legislation and refuses to lead.

In any case, the clean-air bill failed to address what everyone now recognizes to be the overarching environmental problem—global warming. Industrial societies have already produced enough atmospheric contamination to ensure an increase in temperatures in the next few years. Some scientists estimate that if the use of fossil fuels continues at present levels, global temperatures will rise five to fifteen degrees by the middle of the next century. Others believe temperatures will go up three to eight degrees before then. In other words, global climatic conditions are bound to get worse in the coming years even if the government acts right now. The dreadful consequences, from parched farmland in the Midwest to rising tides threatening coastal cities, will steadily get worse as long as government refuses to do anything.

Politicians all acknowledge the dangers of the illness. But they dread the cure—reversing the greenhouse effect will require a profound reordering of industrial societies, for both producers and consumers. Last spring Senator Robert Stafford, Republican of Vermont, pleaded with his fellow senators to confront the implications of global warming. "The first temperature increases," said Stafford,

> are expected to be confirmable beyond dispute in the 1990s, but there are some who believe that the signals of global warming and climate destruction are already manifesting themselves. They cite the fact that four of the last seven years are the hottest on record, that global average temperatures have increased by at least one-half a degree in the last half century, that a wide variety of circumstantial evidence—for example, summer and winter droughts, mid-ocean blooms of algae, death of Caribbean coral—is consistent with the trend of rising temperatures. . . .
>
> The West is in the midst of a two-year drought. Last year forest fires raged throughout the [West]. The ocean's temperatures, as well as its levels, have risen. Icebergs are proliferating in both numbers and size. Whether or not these are the long-awaited signals of the arrival of the hothouse which earth may become, there is no disagreement that unless we change our ways, it will eventually arrive.

The greenhouse effect has many complex causes, but the principal culprit is carbon dioxide (CO_2), a colorless, odorless gas produced by every kind of engine that burns fossil fuels—from automobiles to power plants. Not so long ago, CO_2 was regarded as harmless. Today no scientist disputes the necessity to reduce CO_2 emissions drastically worldwide so as to reverse the warming of the planet. Carbon dioxide, incidentally, is one area of manufacturing where America is still ahead of Japan—the United States is the world's largest producer of CO_2.

The Reagan administration actually made the carbon dioxide problem worse this year, once again gutting the existing federal law on automobile fuel efficiency. Autos are the single largest source of CO_2, and unless people want to give up cars, the only way to reverse the damage they cause to the atmosphere is to insist on automotive engines that burn dramatically less fuel per mile. Reagan's Department of Transportation (DOT) has gone in the opposite direction.

On October 3 [1988], right in the middle of the presidential campaign, DOT announced that it was acceding to demands by General Motors and Ford that it lower the federal standard for fleet averages from 27.5 to 26.5 miles per gallon—a concession that allows more production of larger cars and thus increases atmospheric pollution. In fact, the Reagan administration has let the auto companies off the hook on fuel efficiency every year since 1986—annual relaxations of standards that have added 300 million tons of CO_2 annually to the atmosphere. In a speech to Detroit business leaders this fall, the secretary of transportation, Jim Burnley, denounced the 1975 mileage-efficiency law and called for its repeal. Democrats were silent. Evidently fearful of offending auto workers, the Dukakis campaign passed up what could have been a major issue in 1988—saving our planet from ourselves.

Congress, like the Reagan administration, also made things worse. Marching under the clean-air banner, the Democrats and Republicans enacted a bill to encourage the production of cars that can burn alternative fuels—methanol (made from coal and natural gas) and ethanol (made from grains). These alcohol-based fuels might relieve air pollution on the ground in smog-choked cities like Los Angeles, but for complicated reasons of chemistry, they would not help the larger problem in the atmosphere. In fact, coal-based methanol would actually produce sixty percent more carbon dioxide than gasoline, not to mention ten times more formaldehyde, a cancer-causing chemical.

The methanol legislation virtually guarantees further damage to the atmosphere. It grants the auto companies yet another exemption from the fuel-efficiency law (1.2 miles per gallon on the fleet averages) if they manufacture cars that can burn both gasoline and methanol or ethanol. Of course, once some of these new dual-fuel cars are on the road, the companies will get their exemption and use it to produce more gas guzzlers. Meanwhile, there's nothing to ensure that drivers of the dual-fuel vehicles will actually use methanol or ethanol instead of gasoline.

Clean air, in any case, was not the political motive that pushed the methanol bill to near-unanimous passage. The real driving force was "constituent economics"—currying favor with farmers and coal miners by creating a potential new market for what they produce. A few honest voices like Senator Stafford's were raised in dissent, but their warnings were drowned out. Even major environmental groups belonging to the National Clean Air Coalition ducked the issue. The coalition formally declared its opposition to the methanol bill but concluded, somewhat cynically, that there was nothing to be gained by actually trying to block a measure that was so popular among Capitol Hill politicians. Unfortunately, the politics of the environment is frequently defined by such game playing, parochial trade-offs and short-term tactics.

The battle over strengthening the Clean Air Act this year [1988] demonstrates vividly how a great national issue can degenerate into provincial political feuds. The winners and losers were obvious. West Virginia, Ohio and Kentucky, where high-sulfur coal is mined and burned, defeated Maine, Vermont and New Hampshire, where the lakes and forests are acidifying from sulfurous pollution. Detroit, the capital of autos, shut out Los Angeles, the capital of poisonous air. Wyoming coal won. Denver's asthmatics lost.

On a national scale the really big winners were the electric-power utilities, the auto industry and other heavy manufacturers who always benefit when no new clean-air legislation gets enacted. The immediate losers are the 117 million people who live in areas where air pollution violates health standards prescribed by law. Thirty-five million of these Americans whose air was already below standard have seen it deteriorate further during the last decade.

Senator George Mitchell of Maine, the Democrat who chairs the Subcommittee on Environmental Protection, set out to negotiate his way through the snarl of local interests and reform the Clean Air Act primarily to halt acid rain. (His state, after all, is one of the chief victims of acid-rain pollution from the Midwest.) To get his bill started, Mitchell first had to hand out certain localized exemptions or delays sought by his fellow committee members—special treatment for pollution sources in their home states. Chaffee of Rhode Island, Mikulski of Maryland, Burdick of North Dakota, Warner of Virginia, Graham of Florida, Breaux of Louisiana, all got exceptions and preferential treatment. Diluting clean-air enforcement has become its own form of pork barrel for Capitol Hill politicians—a goody, like federal money for a new dam or highway, to seek for special constituents.

Once his bill was reported from committee, Mitchell turned to the real contest—negotiating terms with fellow Democrats from Eastern coal states and the industrial Midwest. Since Senator Robert Byrd of West Virginia is the Democratic majority leader, everyone understood that until Byrd approved the terms for coal, Mitchell's bill would not be allowed to reach the Senate floor for a vote. Since Mitchell himself is a candidate to succeed Byrd as majority leader next year, he had added personal incentive to work out a compromise acceptable to other Democratic senators.

Why not challenge Byrd head-on and force the issue to a floor vote? Mitchell said the route had been tried in the past and produced nothing. "Senator Byrd had made it clear that even if you offered the legislation as a floor amendment to another bill and you succeeded, he would simply withdraw the bill and you would accomplish nothing," Mitchell explains. "The question is, do you want to make a statement, or do you want to make a law?"

Through the summer months [of 1988], Mitchell bargained repeatedly with Eastern coal—not so much with Senator Byrd but with Richard Trumka, president of the United Mine Workers [UMW]. Meanwhile, he also offered concessions to Senator Alan Simpson, Republican of Wyoming, who represents the Western producers of low-sulfur coal. Pollution control that's bad for West Virginia coal is potentially good for Wyoming's coal, which considers itself a competitor. Given the crass realities of Capitol Hill, Mitchell's back-room strategy was perhaps the only way to proceed.

"The only way to get an acid-rain bill through Congress is to split the opposition—divide the coal miners' union from the electric utilities," explains Leon G. Billings, who served many years as staff director of the Senate Public Works and Environment Committee. "It may not be possible even if you do that. But Mitchell succeeded in doing what for eight years had been undoable—getting the coal miners to accept a deal. Did he give up too much to the UMW? The answer is he gave up what he had to give up to get a deal."

When lobbyists from environmental groups saw the outlines of Mitchell's compromise, they were disgusted. The bill not only reduced the overall mandate for cutting sulfur dioxide emissions—the principal element in acid-rain destruction—but it also pushed significant reductions far into the future, the year 2003 and beyond.

By comparison, the government of West Germany decreed in 1983 that within five years all of its 110 power plants be refitted with the best available technology to eliminate acid-rain pollution. The task was virtually completed this year while the U.S. Congress was choking on a less-demanding plan stretched out over fifteen years. Besides West Germany, other European countries and Japan have also moved aggressively against acid rain. Their forests are dying. So are ours.

When the National Clean Air Coalition declared its opposition to Mitchell's bill, the game was effectively over. The coalition had strong reason to believe that no bill at all was better than what Congress was about to enact. Its logic was this: If the Senate was going to compromise on acid-rain regulation, the House would likely be even more lax on automobile emissions and general air-quality enforcement. And Mitchell's bill, disappointing as it was, would certainly be diluted even further in the closing days of Congress, when the infighting always becomes especially arcane and nasty.

In the House another formidable power, Representative John Dingell, Democrat of Michigan, was geared up to impose his own version of compromise on the legislative game playing. Dingell is the chairman of the House Energy and Commerce Committee, a shrewd and strong-willed player who speaks for his home-state auto industry, as well as for other industrial sectors. Dingell derisively refers to his opponents as "enviros" and "jackass environmentalists."

Year after year, Dingell has succeeded in squelching Representative Henry Waxman, Democrat of California, the chairman of the Commerce Subcommittee on Health and the Environment, whenever he has tried to strengthen the law. This year [1988] Waxman was completely neutralized by Dingell's opposition and couldn't even get the votes to report a bill from his own subcommittee.

"Clean-air legislation would have been passed years ago if Mr. Dingell didn't have problems with Mr. Waxman's legislation," says David Hawkins of the Natural Resources Defense Council. Billings, who now lobbies for the South Coast Air Quality Management District, in Southern California, adds, "You have a guy who is extremely powerful and who uses that power brutally—ask the guys in the House who have tried to go against him—and who is also one of the most effective legislators in the House."

Dingell responds indignantly to the "black hat" label. The "enviros," he claims, promote legislation that won't work and would only damage industry and produce lawsuits. "Who wears the black hat under those circumstances—me or them?" Dingell says. "It isn't just the American automobile industry. It's the steel industry and chemical industry and smelting and electric generation. I don't intend to destitute those people. I think we've got an opportunity to preserve jobs and industry and competitiveness and also improve on human health. I don't think I have a mandate to put one over the other."

Dingell's self-righteous rebuttal is actually a fair summary of why the political system fails to act. Economic interests are now treated with the same respect as those of human health and survival. Perhaps with even more respect. In Washington terms, the environmental debate is now just another argument over money.

The trouble with corporations, says a legislative aide who has struggled for years to enact serious environmental controls, is that they don't have grandchildren. Corporations are legal entities responsible for the next quarter's earnings, not for what happens to the next generation. They create jobs and profits in the here and now. They are incapable of looking over the horizon and accepting responsibility for the future.

If that sounds too harsh, consider E. I. du Pont de Nemours and Company. Du Pont's products have contributed directly to the deterioration of the ozone layer that protects us from the sun's ultraviolet rays. In 1974, when scientists warned that the ozone mantle in the atmosphere was being dissipated by industrial gases such as Freon—du Pont's trade name for chlorofluorocarbons (CFCs)—the company dismissed the evidence as inconclusive. Du Pont magnanimously promised to abandon Freon—used in air conditioning, Styrofoam cups, cushions and the like—if the company was ever presented with proof.

Eleven years later alarmed scientists discovered the now-famous hole that has developed in the ozone layer over Antarctica—a sudden breach in the ecosystem worse than anything the Cassandras of the early seventies had predicted. After some public embarrassment, Du Pont agreed to keep its promise and stop manufacturing Freon. But when? Du Pont will not set a date [as of 1988], but the company says it will be before the year 2000. First it must test possible substitutes for safety. If government were more alert and less compliant, it would have told Du Pont to start testing for conversion years ago.

Meanwhile, politicians around the world, including officials of the Reagan administration, are congratulating themselves on their new international agreement to reduce global production of CFCs by fifty percent by the year 2009. Trouble is, Congress's Office of Technology Assessment estimates that the actual effect of the new agreement will only be a forty percent reduction in CFC pollution—even in the unlikely event that the whole world complies with the accord. No one can say what damage another twenty years of increased pollution will do to the ozone layer, though the world already knows what the last fourteen years of stalling accomplished.

The ozone issue has far graver potential than many other environmental dangers, but the politics is typical. With rare exceptions, the corporate response is first denial, then grudging retreat as the evidence accumulates—accompanied by legal and political maneuvers designed to delay adjustment as long as possible. Lead additives in gasoline—also developed by Du Pont—were known to have been poisoning children for years before the federal government finally, reluctantly, moved against them. Yet the government still allows ten percent of this poison to remain in gasoline rather than disrupt the business of the small refiners who market it. "The environment," says Billings, "is the contact issue between the public interest and corporate America, between organized special interests who control capital and what the public wants and needs."

The auto industry is perhaps the most adept at environmental politics. It held off enactment of meaningful air-pollution controls for more than a decade, then resisted compliance for another decade after the auto-emissions standards were written in law. The controls on cars have been effective in reducing pollutants, but the industry's stall tactics tend to cancel out real progress on overall air quality. While Detroit yielded grudgingly, the number of vehicles on the road increased, and Americans now [1988] drive another 200 billion miles per year—more exhaust, more gross pollution that will require even tougher standards, which, naturally, the industry dismisses as unwarranted.

One obvious solution, which would improve air quality and reverse global warming, is to manufacture automobiles that burn a lot less fuel per mile. The good news is that the necessary technology already exists—in 1985, Toyota unveiled a five-passenger prototype that runs ninety-eight miles on a single gallon of gasoline. It's not yet on the market, however. (Other breakthroughs are possible in other technologies—from light bulbs to power plants—if the federal government would prod both producers and consumers to adopt them.)

The bad news is that auto companies make bigger profits on large cars, especially in America, and so corporate self-interest pushes them in the wrong direction. This will continue, no doubt, until government enacts new taxes that penalize sellers and buyers of inefficient engines or passes laws prohibiting wasteful vehicles altogether.

American auto companies are digging a hole for themselves—and especially for their workers. The automobile manufacturers have shifted most of their production of smaller, more efficient cars to foreign countries. As a result, American auto workers are mostly making larger cars. Their jobs, once the politicians get around to facing the implications of global warming, will be much more threatened than, say, those of auto workers in Japan or Korea. Naturally that only makes the politicians more reluctant to confront the problem.

While corporations are the major cause of the stalemate in the politics of the environment, it's too easy to blame only them. Everyone shares responsibility: For example, the major environmental groups that accumulated political strength twenty years ago when Americans discovered ecology, today seem bogged down in their own version of tired-blood politics. After years of battling in the trenches, fighting polluters in the courts and in regulatory agencies for incremental victories, the environmental professionals seem ground down by Washington's way of doling out progress in small bits and pieces. The movement has lost its anger and radical edge and, more important, its ability to mobilize public opinion into an effective political force.

Washington politicians weren't prepared to do anything this year about the public's anxieties over the global hothouse, but, in fairness, neither were most environmental groups. They have become accustomed to pushing for specific issues on well-defined turf—land, air, water, toxics. The implications of global warming cut across these narrowly defined areas—and will force environmentalists to reexamine their objectives, too. Solutions of the 1970s like catalytic converters on automobiles or scrubbers on power plants actually aggravate the greenhouse problem.

Ultimately, finding answers to our environmental crisis will be much tougher than just blaming political cowardice and corporate greed. The vast adjustments that will be necessary go very deep, to the core of industrial society, and virtually every consumer will be forced to make changes in lifestyle—whether it is giving up those beloved "muscle cars" or accepting that human existence does not require air conditioning twenty-four hours a day.

People got angry this year over the environment, but will they stay angry long enough to be heard in Washington's debate next year? The battle lines are already drawn. Before his retirement this fall, Senator Stafford introduced a far-reaching bill to set standards and deadlines for reversing the formation of the hothouse. Senator Tim Wirth, Democrat of Colorado, and Representative Claudine Schneider, Democrat of Rhode Island, are sponsoring related measures—the first steps in what promises to be a long fight. Meanwhile, a new international network of activists from thirty-five nations was recently formed to arouse public opinion around the world. They plan a Global Earth Day for the spring of 1990 and hope to launch a Green Decade movement to overcome politics as usual.

Jeremy Rifkin, the well-known social critic and president of the Foundation on Economic Trends, believes the Global Greenhouse Network he helped launch this fall must first provoke deep ethical arguments among citizens before the political system can be expected to respond seriously. "The greenhouse crisis is the bill coming due for the Industrial Revolution," Rifkin says. "It's not an accident. It's the logical outcome of our world view—the idea that we can control the forces of nature, that we can have short-term expedient gains without paying for them, that there are no limits to exploitation of the environment, that we can produce and consume faster than nature's ability to replenish."

Perhaps real environmental politics has to begin not in official Washington but with citizens everywhere pondering those somber thoughts.

America's Worst Polluter

Howard Kohn

*T*om Bailie liked most of George Bush's State of the Union speech earlier this year—at least up to the part where Bush promised once again to be "the environmental president."

"I wish I could believe Bush, I really do," says Bailie, a grain and cattle farmer in the state of Washington. "But how can you believe somebody who's in charge of a government that's poisoned our air and water, and then lied and lied about it? I don't mean a little bit of poison, either. The government might as well have murdered us in our sleep."

There's good reason for Bailie's strong words. His farm sits about a mile downwind and downriver from the eastern boundary of the Hanford nuclear-weapons reservation—a stretch known locally as Death Mile. In twenty-seven of the twenty-eight households nearest the Bailie farm, there have been catastrophic health problems associated with radiation:

THE ROLLING STONE
ENVIRONMENTAL READER

thyroid and bone cancer; stillborn births and physical and mental defects in newborn babies; leukemia and other blood diseases; an outbreak of boil-like sores; and sterility. In Bailie's family alone, his four grandparents, his parents, two of his sisters, three uncles and an aunt have died from—or are now suffering from—breast, intestinal or colon cancer. The forty-three-year-old Bailie himself had to spend part of his childhood in an iron lung and is now sterile. "After a while most of us figured the bomb factory had to be the cause," says Bailie.

Over the last five years Bailie and other residents living near Hanford have obtained federal government records confirming their long-held suspicions. Starting in the late 1940s, when Hanford became the country's biggest bomb-manufacturing plant, its smokestacks began spewing radioactive gases, principally plutonium and iodine-131, into the air—in amounts far greater than the leaks produced during the 1979 accident at Three Mile Island. And because the citizens around Hanford weren't warned about the air they were breathing or the milk (from cows feeding on radioactive grasses) they were drinking, the residents were probably hit with a bigger cumulative radioactive wallop than anything Chernobyl residents experienced. "We were zapped," Bailie says bitterly.

The dirty secret of Hanford, unfortunately, is part of a much larger scandal. The U.S. government, which many Americans assume is faithfully working to safeguard their environment, instead has been the nation's single worst polluter for the last forty years.

Pollution attributable to federal departments and agencies bombards the air, land and water every day, in every region of the nation, and in a half dozen foreign countries as well. The Department of Defense (DOD) and the Department of Energy (DOE) do the most damage, but virtually every department and agency under federal control is an environmental offender. Far-flung and diverse, our government's environmental abuses include mishandling of radioactive and chemical wastes that are leaching into water and soil; use of unnecessary toxic materials that produce not only hazardous wastes but also destructive atmospheric gases and acid rain; neglect of public lands to the extent that some of them have become monstrous waste dumps; and pursuit of policies that have led to the ruin of wilderness areas and the erosion of huge tracts of farmland.

Bob Alvarez, a senior investigator for the U.S. Senate Governmental Affairs Committee, has for fifteen years been documenting the government's sorry environmental record. "What kind of example is it when the government has done more to destroy our environment and risk our health than anyone else?" says Alvarez. "How can the government in good conscience enforce the law against Exxon and the other corporate polluters when it is de facto the biggest outlaw?"

The U.S. government's environmental lawlessness is substantiated by its own documents. According to an astonishing report issued last year by the General Accounting Office, federal departments and agencies have been collectively violating clean-water laws at twice the rate of private industry. A 1988 survey by the Environmental Protection Agency (EPA) found that fifty percent of the facilities under federal jurisdiction were guilty of causing some form of environmental damage. "The federal government is not obeying its own laws," says Alvarez. "It has adopted the attitude that it is above the law."

The EPA, first established as a federal agency shortly after the original Earth Day, in 1970, was empowered to impose civil fines and criminal penalties on polluters. Yet while the agency has had some success cracking down on cities, states and private companies for failure to comply with environmental laws, it has been almost completely ineffective when the federal government itself is the culprit. "No question there's been a double standard," says Representative Mike Synar, the Oklahoma Democrat who chairs the House Government Operations Committee's subcommittee on the environment. "The federal government has been inexcusably soft on itself."

Part of the blame for this goes to Congress, for allowing the defense and energy departments to claim exemptions from environmental laws on the grounds of national security. But more important is an explicit White House policy—set in place by Ronald Reagan and continuing even now under George Bush—to block the EPA from assessing penalties against federal departments and agencies. That policy was part and parcel of Reagan's drive to deregulate the U.S. government in general and to pass the costs on to future generations. The man who headed the Reagan administration's deregulatory task force was Vice President

Bush. "It's a terrible irony that during a time of world peace the U.S. government has waged what amounts to a war of pollution against its own people," says Mike Clark, president of Friends of the Earth, a major environmental group.

Billions of pounds of the government's accumulated waste—containing plutonium, arsenic, cyanide, nerve gases, TNT, rocket fuels, pesticides, bacteria and other harmful substances—pose the most immediate threat to public health. Most of these hazardous wastes are generated by the DOE's and DOD's nuclear- and conventional-weapons factories, which are for the most part operated by private contractors under federal supervision. But significant amounts of toxic waste also emanate from NASA's space program, Department of Agriculture research, mining on Department of Interior lands and government-wide equipment maintenance. They are often carelessly discarded in the open environment or haphazardly stockpiled, creating thousands of radioactive and chemical time bombs. "We'll be paying the price for generations," says Clark.

Storing hazardous wastes has become a particularly urgent problem. Because the government has made only the barest low-cost effort to provide adequate containment equipment, tons of toxic materials have spilled out or seeped into scores of bodies of water—some of which are connected to drinking faucets in American homes. Waters leading to drinking supplies in and around Denver, Tucson, Cincinnati and Portland, Oregon, due to their proximity to weapons factories or munitions plants, are threatened with radioactive or chemical contamination. Seven states have already closed more than 300 drinking wells they deemed unsafe.

Another major problem is the government's habitual use of destructive chemicals for routine cleaning and maintenance of its computers, vehicles and other equipment. Instead of soap and water, federal agencies annually use thousands of gallons of solvents containing chlorofluorocarbons (CFCs), which scientists say eat away at the earth's ozone layer, allowing the penetration of deadly ultraviolet rays. With the world's largest fleet of cars, trucks and armored vehicles—many of them crude gas guzzlers—the government is also the single biggest contributor to dirty air, as well as to the greenhouse effect.

Further long-term environmental degradation is being caused by government policies that continue to open up public lands to commercial exploitation, putting private profit ahead of environmental protection. While these policies date back forty or fifty years they became a kind of religion under former president Reagan and his secretary of the interior James Watt. The U.S. Forest Service, for example, continues to encourage and subsidize widespread cutting of ancient, publicly owned trees—thereby removing one of the most effective antidotes to global warming. Every year timber companies chain-saw another 60,000 acres—three times more than during the construction boom of the 1950s, when logging of the national forests first began in earnest. Perverting its original mission, the Forest Service now spends more money cutting down trees than protecting them.

The government has betrayed its responsibilities to public lands in other ways. Federal policy, by actively encouraging oil and gas drilling in wilderness areas, has destroyed countless wildlife habitats. Meanwhile, federal negligence allows mining companies to leave behind mountains of toxic residue in national parks, and municipalities to deposit garbage containing hazardous waste onto 1,000 landfills on federal land.

Another federal policy, advocating massive overplowing of farmlands, has caused more soil erosion in the last two decades than during the bleak Dust Bowl era of the 1930s. The runoff from farms, laden with silt, manure or high-powered pesticides and fertilizers, is now suffocating the delta area of the Gulf of Mexico, the Chesapeake Bay and Lake Okeechobee, in central Florida, among other major bodies of water.

As Shira Flax, a Sierra Club toxics expert in Washington, D.C., puts it, "Twenty years after the original Earth Day, there's one great unregulated polluter left in this country—our own government."

By far the most dangerous form of pollution caused by government mismanagement comes out of the Department of Energy's nuclear-weapons facilities—a total of seventeen factories, laboratories and test sites scattered around the country. After four decades of research and manufacturing, the DOE is stuck with billions of gallons of radioactive and chemical by-products, principally at four locations: the Hanford facility, in Washington; the Savannah River tritium plant, in South Carolina; the Rocky Flats plutonium plant, in Colorado; and the Fernald uranium plant, in Ohio. "At Hanford alone, the wastes would cover an area the size of Manhattan with a lake forty feet deep," says Senate investigator Alvarez.

Perhaps the most acute problem lies in the makeshift ways these plants store their wastes—either in open, unlined ponds or in deteriorating underground steel tanks. For years the DOE and its predecessor, the Atomic Energy Commission, assured people living in communities near the bomb plants that the ponds and tanks were providing safe containment. But tests conducted over the last ten years at all four facilities showed conclusively that dangerous amounts of radiation were leaching and spilling out of the ponds and tanks—many of them cracked and corroded—into the local water systems.

At Savannah River, for example, DOE scientists had confidently predicted since the 1960s that it would take at least 1 million years for plutonium from the plant's stored radioactive wastes to reach the water table. But in the late seventies, significant traces of the radioactive isotope began to be detected in underground wells. When test readings were kept hidden from the public, Bill Lawless, then a DOE senior engineer at Savannah River, blew the whistle and ultimately testified before Congress. "Basically, the bureaucracy drowned the truth," Lawless says now. Today radioactivity has been found in Four Mile Creek, which flows into the Savannah River, the region's biggest water source.

DOE officials had also maintained there were no serious storage or discharge problems at the Fernald plant, located nineteen miles from Cincinnati. But last year a group of local residents won a settlement of its class-action suit after, among other things, it was discovered that eighty-one tons of liquid uranium had entered the Great Miami River, a major freshwater body in the Cincinnati metropolitan area. And in spite of repeated DOE assurances, radioactive wastes from the Hanford plant were also found in the Columbia River, which supplies water to communities from eastern Washington to western Oregon; similarly, the main water supply for some suburbs of Denver has been contaminated with chemical and radioactive wastes from Rocky Flats.

The DOE has also failed to keep its nuclear-weapons plants from emitting radiation pollution into the air. The cause of this is equal parts defective technology and officially sanctioned negligence. At Fernald, the DOE was forced to acknowledge that at least 197 tons of uranium dust had escaped over the years through the ventilation system as much because of poor supervision as faulty dust-collector bags and gummed-up scrubbers.

"In essence, the DOE fabricated scientific data for thirteen years," says Arjun Makhijani, president of the Institute for Energy and Environmental Research, who examined Fernald's records on behalf of the local residents. Makhijani found long rows of zeros blanketing the documents, suggesting that the filters had operated perfectly. Upon further checking, however, he discovered that the zeros meant that for months on end no one had taken any measurements at all. In the meantime, radioactive particles were, in fact, spewing out the vents.

"It's still difficult to get a totally accurate picture, because for so long the DOE took the position that everything it did was a state secret," says Representative Dennis Eckart, an Ohio Democrat. Most of the information about the DOE's terrible pollution history has been forced out into the open by an avalanche of lawsuits and state and congressional investigations. Assessments of the consequences to public health have finally gotten under way—but in a piecemeal fashion. Today the only independent studies in the Hanford and Fernald areas are being conducted by the Centers for Disease Control. They began a year ago with no results expected for at least another few years.

"The story of the bomb plants," says Senate investigator Alvarez, "is that they're just waking up after a forty-year binge of thinking there's no tomorrow. Well, it's tomorrow, and welcome to the hangover."

The one federal department that ranks with the DOE as an environmental polluter is the Department of Defense. None of its approximately 2,000 facilities, which include military bases, munitions factories,

training grounds, research laboratories and radar sites, produces anything quite as deadly as radiation. Yet the Pentagon is responsible for a
wide variety of environmental problems, foremost among them the
improper handling of chemical wastes.

Operators of an Air Force missile factory outside Tucson, for example,
regularly discarded paint, crankcase oil and solvents containing cancer-
causing trichloroethylene (TCE) by dumping them into a ditch and other
open pits. Residents of one Tucson neighborhood, Southside, have
charged in a lawsuit that because of this careless dumping, their drinking wells have been tainted with TCE—at forty times the permissible
level. "The whole neighborhood was poisoned," says one resident.

These dumping practices, unfortunately, are standard operating pro-
cedure at defense installations around the country. At the Army's Corn-
husker ammunition plant, in Nebraska, a seeping swath of chemicals
from the manufacture of TNT has ruined 300 private wells in the
suburbs of Grand Island. In Maine, state officials have concluded it is
unsafe to eat shellfish caught near the Portsmouth Naval Shipyard because of lead and toxic PCBs dumped there. At the Army's Rocky
Mountain Arsenal, in Colorado—once the country's biggest manufacturer of chemical weapons—discarded nerve gases, mustard gas, lead,
mercury and arsenic have infiltrated thirty square miles of
groundwater—apparently causing the yellowing of crops, the death of
livestock and various cancers and blood and skin disorders. In Minnesota, wells in an eighteen-square-mile area around the Army's New
Brighton munitions facility are contaminated with TCE, and west of
New York City, an underground stream containing TCE 5,000 times
above permissible levels is flowing from the Army's Picatinny Arsenal
toward a planned subdivision of 1,600 homes.

The Pentagon, which is supposed to be responsible for cleaning up its
own hazardous wastes, has not even identified all its problems. Inspections at 1,579 facilities have revealed 14,401 sites with threatening toxic
wastes. To date, only about two percent have been cleaned up.

Some of these hazardous wastes are a legacy of sloppiness and bad
management dating back twenty or thirty years. But in many cases there
is also a pervasive arrogance among Defense Department officials that

makes them utterly disdainful of the environmental consequences of their operations. Officials at the Army's Aberdeen Proving Grounds, in Maryland, had ignored repeated warnings about a deteriorating containment dike. It finally broke in 1985, spilling hundreds of gallons of sulfuric acid used in nerve-gas production into a creek that feeds Chesapeake Bay.

Even when its guilt is obvious, the Pentagon often refuses to make amends. For several years residents of a community outside Jacksonville, Florida, noticed a steady convoy of trucks ferrying wastes from the Jacksonville Naval Air Station to a local landfill. When local wells began to test positive for toxic chemicals and some residents started contracting immunity disorders, rheumatoid arthritis and other illnesses, Navy officials repeatedly denied they were dumping at the landfill. The officials had to backtrack after residents dug six inches into the landfill and exposed containers bearing the Navy logo. Yet, to this day, the Navy has failed to compensate any of the victims. "I thought the government was supposed to look out for its citizens," says Yvonne Woodman, a resident who contracted lupus, a sometimes deadly blood disease, after drinking the local water. "But all the Navy's done is look out for itself."

For sheer haughtiness, however, nothing can top this common scam: When Pentagon officials auction used equipment by lot, they will often dump acids, solvents and other dangerous chemicals in with the valuable hardware. The toxins get pawned off on military-surplus dealers, who discard them wherever they can. The Los Angeles police last year discovered 50,000 gallons of military chemicals, highly corrosive and explosive, that had been left in a vacant lot under a freeway, just a few feet from a river.

There is hardly a major environmental problem to which the Pentagon does not make a substantial contribution. Every branch of the military, with the exception of the Air Force, still insists on cleaning much of its equipment with CFC solvents, even though they are being phased out elsewhere in the United States and in other developed countries. Military vehicles, especially jets on training missions, also exacerbate such problems as air pollution and acid rain. The jets burn an estimated 8,000 gallons of kerosene-based fuel per hour, five percent of which comes out the exhaust valves as an acid-rain-producing mixture of carbon monoxide, nitric oxides, hydrocarbons, soot and sulfur dioxide. And a GAO [General Accounting Office] report last year singled out the Pentagon's extensive training flights and ground exercises as a cause of damage to the nation's 452 wildlife refuges. The cold war may be over, but Pentagon officials still want to play their war games. Over the objections of environmentalists and ranchers out West, the DOD is going full speed ahead with a plan to add another 4.6 million acres of wilderness to the 25 million it already has for staging training exercises.

While not to the same extent as the DOE and the DOD, a dozen other departments and agencies under federal control also have appalling environmental track records. What they have in common is chronic ineptitude in the handling of sewage, toxic chemicals and other hazardous substances. Among the offenders are NASA (for dumping rocket fuel or other chemicals at twelve facilities, including the Jet Propulsion Laboratory, in Pasadena, California, and the White Sands test center, in New Mexico); the Federal Aviation Administration (for discarding solvents and waste oils at fifty-three airfields in Alaska); the Department of Agriculture (for dumping pesticides and other chemicals at twenty-seven research labs and for spraying the fumigant carbon tetrachloride on surplus grain in 2,000 silos in the Farm Belt); the Coast Guard (for jettisoning toxic waste into lagoons, fire-training pits or the open ground at eight bases) and the Justice Department (for mishandling sewage at seven federal prisons).

As head of the EPA's Federal Facilities Hazardous Waste Compliance Office, Christopher Grundler is the man in charge of the overall effort to root out toxic-waste polluters. "All around the country, you can find little polluters being hit over the head, while right around the corner a big polluter goes free," says Grundler. "And the big polluter is the federal government. It's more than ironic: It's completely unfair."

How does the government get away with it? The problem goes back to the ambiguous authority the EPA has had from the beginning to enforce environmental laws against other federal agencies. Taking advantage of the legal muddle, Ronald Reagan and his attorney general Ed Meese adopted the position that departments and agencies of the federal government could not be prosecuted for environmental violations. In 1986, Reagan and Meese formalized this philosophy into a legal theory called "unitary theory of the executive," which effectively stripped the EPA of all enforcement powers at the federal level.

During the 1988 presidential campaign, George Bush himself conceded the federal government's poor record on environmental self-policing. "Unfortunately, some of the worst violators are our own federal facilities," Bush said in a Seattle speech. "As president, I will insist that in the future federal agencies meet or exceed environmental standards. The government should live within the laws it imposes on others."

Yet so far little has changed. A White House spokesman explains that the issue of EPA authority is under review. But EPA officials, now under the strong leadership of William Reilly, are growing impatient. "With a stroke of the pen, Bush could cancel Reagan's policies and start to untie the EPA's hands," says one EPA man. "But Bush's Justice Department is applying the same legal theories as Reagan's."

Another of those theories holds that federal departments and agencies are also immune from state enforcement. Whenever state environmental officials have stepped in and tried to impose penalties on federal violators, the Justice Department has moved to quash them. "Whether the federal government can be bound by the states is an issue for the courts to decide," says the Bush spokesman.

And with its limited prosecutorial power, the EPA can do little more than jawbone. "Since we can't levy fines, we're constantly trying to figure out new ways to say *please*," says Grundler. When the EPA set a 1988 deadline for the Pentagon to install a safer waste-treatment system at Rocky Flats, energy officials challenged the deadline in court, setting off a protracted legal battle. An efficient treatment system has yet to be put in.

The EPA did declare a victory when it got the Army and Shell Oil, the two managers of the Rocky Mountain Arsenal, to agree to a voluntary cleanup. But the agreement was not far-reaching enough to satisfy Colorado governor Roy Romer. He directed his state attorney general to sue the Army, Shell and the EPA itself for the right to oversee part of the cleanup. In a breakthrough ruling, U.S. District Court judge James Carrigan last year granted Colorado partial jurisdiction. In strong language he denounced the EPA deal as one "between two polluters, the army and Shell . . . [to] limit cleanup standards, thus lowering costs for both defendants."

Officials and citizens in several states have devised other strategies for putting the squeeze on the federal government. They have even gone to federal court to file civil suits against the DOE and the DOD, seeking restitution for the enormous cleanup costs and other damages. "There's a new Boston Tea Party in the works," says Alvarez. "People at the state and local level are rising up in open rebellion."

More and more of late, the government has been finding itself a defendant in environmental actions in every region of the country. In the lawsuit brought by neighbors of the Fernald plant, a federal judge ordered a $73 million settlement to the plaintiffs last year for their emotional distress and loss of property values. Meanwhile, the state of Ohio has filed suit against the DOE for nuclear contamination at its Piketon facility; the state of Colorado is seeking compensation for damages from the Rocky Mountain Arsenal; the state of Minnesota is contemplating a suit against the Twin Cities Air Force Reserve Base over chemical wastes; and residents of Tucson, Jacksonville and other cities where drinking wells have been contaminated by military waste are gearing up for court battles as well.

The man most in the hot seat right now is George Bush's new Department of Energy secretary, retired admiral James Watkins. Responding to a fusillade of bad publicity triggered by protests from irate citizens, national environmental groups and a few outspoken members of Congress, Watkins is taking an entirely different tack than his predecessors did. In speeches and congressional testimony, he has made what amounts to public confessions of past abuses within his department, characterizing the DOE as "careless" and "unaccountable." He has pledged to elevate the status of environmental concerns in future policies and decision making.

In the last eighteen months, the DOE has sharply curtailed production at its nuclear-weapons facilities, ordering temporary halts at Rocky Flats, Hanford and Savannah River and preparing for a permanent shutdown at Fernald. But even that will not solve all of the DOE's problems. Barring the unlikely event of total disarmament, the bomb plants will continue to churn out weapons and to produce some amount of radioactive waste. And even if there is total disarmament, the accumulated stored waste already in place will continue to constitute a threat. Watkins is well aware that at Hanford, for example, if the wastes stored in several of the 149 tanks are not adequately stabilized and removed, they might well explode from their internal combustible mixture of cyanide, nitrates and nitrites. Such an explosion was recently revealed by Soviet officials to have taken place in 1957 in the U.S.S.R., turning the homeland of Ural Mountain villagers into a no-man's-land.

A long-standing DOE plan to move stored wastes to a man-made cavern 2,100 feet below the New Mexico desert has been put on hold indefinitely. Geologic pressure on the cavern has turned out to be twice the level originally estimated, and cracks are showing up in the ceiling. Watkins and his staff may have to go back to the drawing board for a long-term solution to their storage problems.

A further challenge for Watkins is how to get control of a bureaucracy accustomed to suppressing information. DOE officials had known for years about the explosive conditions of the Hanford waste tanks from a study conducted in 1984. But the report was kept secret until last October when Senator John Glenn obtained it and released a copy.

"Old habits die hard," says Representative Synar. "Change is coming, but it's not coming fast. Throughout the federal government, the approach is still 'Go slow.'"

Indeed, the foot-dragging extends to one pivotal question: How much will it cost the federal government to clean up its mess? Researchers for the Congressional Budget Office say it's impossible to get a fix on total costs right now; there are still a number of departments and agencies, including the Pentagon and the Department of the Interior, that have yet to come up with a serious plan to address their pollution problems. One independent estimate of the eventual cleanup costs for the defense and energy departments together comes to a staggering $330 billion—twice the officially estimated tab for the bailout of the savings and loan industry.

For the next fiscal year, Watkins has requested $2.8 billion to begin the process of cleaning up—a figure that's getting mixed reviews in Congress. But the Pentagon request is for only $500 million, half the amount many would consider a bare minimum. Defending the low figure, Pentagon officials say that as the cold war dies down and military bases are closed, they want to seal off the most contaminated facilities with fences rather than clean them up. To environmentalists this is no solution at all.

How willing is the federal government to get tough on itself? The answer may be reflected in recent revelations that at one time five of the EPA's thirteen research laboratories required substantial remodeling. These labs had been cited by inspectors—some of them from the agency itself—for violating the very environmental regulations the EPA is charged with enforcing.

Whitewash: Is Congress Conning Us on Clean Air?

William Greider

Although Earth Day provided everyone with a righteous high about saving the planet, dirty old reality mocks the earnestness of that commitment. While millions of Americans were swarming to green rallies across the nation, signing pledges and planting trees, the environmental movement was getting rolled in the real world of Washington politics.

The reality, as environmental lobbyists in Washington will concede, is that the new clean-air legislation that is now heading toward passage—the first to be enacted in more than a decade—is pitifully little, considering the public sentiment for action. Its major regulatory provisions are strewn with deceitful gimmicks and loopholes that will allow industrial polluters to stall and evade for years to come. Above all, the new law does not challenge the wasteful habits built into the American economy.

In some crucial aspects, the new clean-air law will actually be weaker than the old law, which the federal government didn't do much to enforce anyway. This time, the government is retreating, quite explicitly, from the general promise it made to citizens in the first clean-air bill, passed in 1970—that human health deserves a higher priority than private economic interests.

Though the legislative action is not over yet, it is already clear that the big winner in this political fight will be industry—especially auto, steel and chemical companies—which once again will dodge an aroused public opinion. These special interests were helped, of course, by the "environmental president," who denounces mainstream environmental groups as "extremist." But industry was also assisted by Senator George Mitchell, the Senate majority leader, and other Democratic congressional leaders who decided to collaborate rather than fight.

To begin with the positive, the legislation does include one major achievement—the first federal regulation aimed at curbing the acid-rain destruction of lakes and forests. The bill will impose on coal-burning public utilities a ten-year deadline to reduce their sulfurous pollution significantly. One could argue over the timetables and limits, but this is a real breakthrough in a political battle that environmentalists started a decade ago.

A reduction of 10 million tons of sulfur dioxide in the air will slow the destruction of nature as well as improve human health. On the other hand, the measure also creates "pollution rights" for industries so that those who reduce emissions beyond federal standards can sell their excess margin to another industrial polluter who is in violation. This approach creates a marketplace where quantities of air pollution will be bought and sold like a valuable commodity. Once bestowed, this privilege will someday come back to haunt the environmental movement; if private industry owns the right to trade in pollution, it will be doubly difficult and costly to push for tougher standards.

Beyond the acid-rain provisions, the progress in this bill is illusory. In general, the 1990 legislation simply rolls over the old deadlines that were enacted twenty years ago. The first Earth Day led to a legislative declaration that by 1975, Americans everywhere would be breathing healthy air. Congress twice extended the original deadline. In 1977, the law was amended to set a new deadline—1987.

In important respects, the air is cleaner today [1990] than it was twenty years ago—but it is still, by and large, unhealthy. One hundred cities are in violation of the federal standards, and more than 150 million citizens are breathing bad air. In thousands of locations, citizens are still being actively poisoned each day by breathing their neighborhood air. The new legislation simply gives everyone new deadlines—America's most polluted city, Los Angeles, gets another twenty years. "If you live in L.A. and you have a baby this year, he or she will be an adult before the air they breathe meets the national health standard," says Bill Klinefelter, a lobbyist for the National Wildlife Federation. That assumes, of course, that the new deadlines will be taken more seriously than the old ones.

Moreover, the Senate has passed a version of the bill that actually weakens the federal government's ability to enforce these deadlines. The original law created the Environmental Protection Agency (EPA), giving it a legal mandate to step in and impose clean-air plans on states and municipalities that were hopelessly negligent.

This year [1990], at the urging of the EPA itself, the Senate repealed the federal authority to intervene—equivalent to a policeman asking to be relieved of his weapon. Environmental lobbyists are still trying to restore the provision, because the threat of federal intervention provides leverage and gives political cover to local officials who take tough action against industries. The Senate bill sets up instead a cumbersome process of annual penalties—a dubious proposition, considering the EPA's past timidity.

In fact, the worst aspect of this change is that it effectively cuts off the ability of citizen groups to file lawsuits that could force the EPA to uphold the law. Over twenty years [1970 to 1990], the EPA's most forceful actions often came about after the Natural Resources Defense Council (NRDC) or other citizen groups sued the federal agency itself for not complying with its legal obligation. The easiest way to shut up these meddlesome citizens is to erase the EPA's mandate.

"You need the mandatory federal language so citizens can enforce the law when the EPA won't do anything," says Richard Ayres, the NRDC official who heads the Clean Air Coalition, which is dominated by the so-called Group of Ten, a consortium of major environmental and labor organizations.

And though the biggest source of air pollution is automobiles, the auto industry basically gets a free ride under the new federal legislation. Congress is merely adopting auto-emission standards that have already been enacted by the state of California and are enforceable by 1994. Any company intending to sell its cars in California, the nation's largest market, is already planning to comply with the tougher requirements.

Furthermore, various forward-looking proposals that would have pushed Detroit and foreign-car manufacturers to develop new emission-control technologies and more-efficient automobiles—breakthroughs that would help reduce the threat of global warming—were scuttled in the private deal-making among Senate Democrats, the White House and industry lobbyists. The issue of global warming may excite popular opinion, but it is not yet real for Washington.

"This bill isn't pushing anything forward," Ayres says. "In no place does Congress have the nerve to say to industry what it said in 1970: 'Public health requires this. You say you can't, but we're telling you you have to.' Instead, a cautious Congress says, 'We won't push you beyond what you already do.' "

The federal government's dereliction of duty is most shocking in the realm of regulating air toxics, the hundreds of poisonous chemicals that become airborne through various industrial processes. Many of these substances are known to cause cancer; all of them, in sufficient concentrations, make people sick or kill them. The 1970 law ordered the EPA to regulate these industrial emissions. In twenty years of stalling, dodging and fighting off court orders, the EPA has managed to issue regulatory standards for a total of seven toxics.

This is not a matter of scientific uncertainty. The cancer epidemics surrounding refineries and petrochemical plants are now well documented. Indeed, under prodding from Representative Henry Waxman of California, the EPA this year [1990] issued a list of 149 potential high-risk factories, mills and refineries that poison their localities with emissions of butadiene, chloroform and other deadly gases. According to the EPA, each of these plants produces for its neighbors a cancer risk that is greater than 1 in 10,000. By comparison, public-health policy asserts that the acceptable risk should not exceed 1 in 1 million.

Fifty-two of these plants have a cancer risk exceeding 1 in 1,000. Six of them—smelters and chemical-processing plants operated by Asarco, Mobil, Goodyear, Uniroyal, Shell Oil and American Chrome—have a risk higher than 1 in 100. The most deadly of all is the Port Neches, Texas, refinery operated by Texaco. Its emissions of butadiene yield a cancer risk rated at greater than 1 in 10.

Texaco is among the 150 companies that recently signed the Chemical Manufacturers Association's [CMA] double-page Earth Day ad in the *Washington Post.* The ad proclaimed the CMA's new "Responsible Care initiative" to save the planet for the next generation. As it happens, the CMA's list of corporate signers mirrors the EPA's list of high-risk producers. The chemical divisions of Du Pont, Dow, Exxon, Weyerhaeuser, BFGoodrich, W. R. Grace and many others show up on both lists.

The chemical companies, having successfully resisted enforcement for twenty years, recognize that they now have a PR problem; their pledges are designed to soothe the public. But they also have a political agenda—fashioning a clean-air bill they can live with comfortably for many years. They appear to have succeeded.

After two decades of inaction on air toxics, frustrated environmentalists are ready to strike what one of them admits is a "devil's bargain." They are backing off from the health-first standard enunciated in the original law in exchange for immediate action by the EPA to regulate emissions on a specified list of 191 toxic substances. Unfortunately, the

EPA's new standards—assuming the EPA actually complies with the law—will not be based on health risks but on the best existing control technologies already in use by industry.

In other words, the government is asking everyone to tighten up his plumbing—technical fixes that refineries, smelters and chemical plants could have undertaken years ago. Then, many years hence, the EPA is supposed to reexamine the results and decide whether there is still any "residual risk" to human health that needs to be addressed by further regulation. Ayres and other strategists believe significant reductions will be achieved quickly under this approach but concede that the health standard itself has been gutted.

Given the tangle of sliding timetables and elaborate exceptions written into the law, it will be ten to twenty years before the EPA would even have to consider imposing regulatory controls based on health standards. Since the air-toxics provision is loaded with the kind of fuzzy words that lawyers love to litigate, a company could spend the next decade or so in court challenging EPA regulations on how much it has to tighten its plumbing.

Until the technology phase of the new clean-air bill is resolved, citizens will not be able to invoke the health standard as the basis of their legal complaints. That means the pace of compliance will be largely dependent on the decisions of industry, not the power of government enforcement. A decade from now, the public will find out whether the chemical manufacturers were sincere. If not, there will no doubt be another clean-air bill to clean up this one.

"What we wanted," says Ayres, "is a health-based standard—1-in-1-million cancer risk. The Senate bill still has the requirement, but there are forty pages of extensions and exceptions and qualifications and loopholes that largely render the health standard a nullity."

One of the grossest loopholes was won by the steel industry, whose coke ovens are among the most deadly sources of pollution for both workers and communities. Coke ovens should have been on the list of targets twenty years ago when clean-air legislation was first debated. Nothing happened. According to the EPA, there are now twenty-six coke ovens that pose a cancer risk greater than one in 1,000 and six where the risk is greater than 1 in 100.

Yet the new clean-air bill will give the steel industry another thirty years to deal with the problem. John Sheehan of the United Steelworkers of America, who has lobbied for two decades to clean up the coke ovens, calls it the "steel fix."

Yet there are some provisions even worse than the steel fix. After fifteen years of resistance, Du Pont and other manufacturers of chlorofluorocarbons (CFCs)—the ozone-depleting gases—now accept the necessity of phasing out the most damaging CFCs in the next ten years. But the companies plan to substitute alternative gases known as hydrochlorofluorocarbons (HCFCs), which, while much less dangerous, will still eat away at the atmosphere's protective mantle of ozone. The bill codifies the transition already underway.

Since the CFC substitutes are still going to exacerbate the ozone problem, the tough political question facing Congress was this: How long should industry be allowed to go with this unsatisfactory solution? In the Senate bill, the answer is forty years. Du Pont will be permitted to manufacture HCFCs until the year 2030. Du Pont's lobbyists may be disappointed, however, because they were holding out for a deadline of 2060.

As environmental activist Richard Grossman observes, the worst aspect of the clean-air legislation is that its passage in 1990 will effectively close the subject for the next five or ten years. While none of the fundamental changes proposed by Earth Day orators are being initiated, President Bush will undoubtedly wave this law like a campaign banner for the rest of his presidency. Democrats in Congress, having served up this mush in cooperation with the president, are likewise going to claim that it is state-of-the-art legislation.

In the meantime, serious environmentalists should begin asking some hard questions. Ayres blames campaign contributions from industry for thwarting the public's hopes in Congress this year. "The goddamn system has been bought—Democrats and Republicans," he says.

Political money is certainly a big part of the explanation, but it is the easiest part to confront. Another reason industry is able to win so many of the close-in legislative fights—and stall law enforcement later in the executive branch—is economic blackmail. Its propaganda skillfully steers the debate into a crude choice between jobs versus clean air. This year one trade organization papered Capitol Hill with a jobs-at-risk map. Though the statistics were outrageously hyped, this is scary stuff for politicians, and the propaganda was not effectively refuted.

The political reality that enviro advocates are sometimes reluctant to face is the economic implications of their ecological values. While industry pumped out inflated estimates of the costs of this new legislation, the fact is that a genuine clean-air law would probably cost much more than even industry claimed. Real pollution control would mean fundamental changes in production processes and flat prohibition of some dangerous chemicals—and inevitably, the dislocation of some industrial workers.

A serious environmental movement would begin talking about the real costs of change, including the public's obligation to aid workers displaced by reforms. Sketching out the future in Earth Day speeches enlarges everyone's vision of the possible, but nervous politicians require a practical blueprint of how to get there.

A serious environmental movement would also begin asking itself how to develop the political muscle necessary to punish politicians who oppose its agenda. As of now, elected officials sense that they can have their cake and eat it too. They can enact hollow environmental laws that appease public opinion and at the same time take care of their industrial patrons by writing in protective loopholes. As of now, I think the politicians have got it right.

PART III

Rain Forests

*P*ublic opinion has been mobilized. Research has been carried out. Money has been raised. But the cutting and burning of forests—both rain forests and North American old-growth forests—continue unabated, almost uninhibited.

Tens of thousands of square kilometers of forest are lost every year. And tens of thousands of species disappear along with them. The numbers are not precisely known: not the number of species lost, nor even the number of species that had existed.

Forests offer multiple benefits to societies and economies. They regulate regional and global climates by absorbing carbon dioxide. They maintain water cycles and store heat and cold. They provide a habitat for species of all types. They anchor soil, prevent erosion and the silting of rivers. They also yield crops, which can be harvested for consumption or profit. Science draws from forests a supply of genetic materials for pharmaceuticals, industrial chemicals and other products. And forests are among the most impressive of the world's resources, engaging artists and touching the human imagination.

A forest is a place ripe for exploitation; it is also a home, a habitat, an area of incredible beauty. Over the last few decades, however, forests also have become a battleground. On these lands rich with life, the forces of those who wish to preserve them and those who wish to clear them clash with sometimes deadly ferocity. The articles that follow

attempt to portray the multiple facets of this complex issue, which has produced some of the most heated debate,
both internationally and nationally, in modern times.

So what causes the destruction of forests on such a massive scale? One answer is poverty. Some of the poorest
regions of the world are home to its richest forests. In a
desperate attempt to survive and feed their families, the
poor of the world clear acres of rain forest to farm the soil.
Another answer is greed. Bill McKibben writes of the economic imperatives that have overtaken conservation efforts
for America's forests.

Deforestation has been exacerbated in the Southern
hemisphere, some say encouraged, by misguided national
and international policies. Rather than work to protect rain
forests as a sort of "environmental capital" that would help
to support productive economies for years into the future,
governments and development banks have chosen to use
them up and forgo their long-term benefits. The reality of
rain forest politics is that some people pay the costs of forest loss, while other people take the benefits.

Even though public opinion may oppose the cutting of
forests and research may show that optimal use of the forests requires their protection, there are still individuals
who will benefit from exploitation for short-term profit. So
far, few plans for protecting forests have addressed the social bases and legal and economic structures that reenforce
their destruction. And these may be the most difficult
topics to address of all.

Recall of the Wild

Winifred Gallagher

"Welcome to a tropical forest," hollers Dan Janzen. His outstretched arms extend toward a desolate landscape of weeds, volcanic rock and gravel shimmering in the heat of high noon. Just behind him, big trailer rigs roar down the Pan-American Highway from Nicaragua. Power lines bisect the sky. The place has the natural beauty of the industrial outskirts of Newark, New Jersey. But Janzen wraps his head in a grubby cloth bag he otherwise uses for toting snakes and starts snapping away with his Nikon like a tourist at the Grand Canyon. "This was a landfill pit," he says, "but in twenty years I'll take this same picture and there'll be so many big trees you won't be able to see the road."

This pockmark in the highlands of Costa Rica's northwestern coast is part of the two-year-old Guanacaste National Park (GNP). The park's

231-square-mile patchwork of previously conserved woodlands and recently acquired ranches, rivers, mountains and beaches is already a living
museum of one of earth's last wild corners. This melancholy, if laudable,
function doesn't satisfy Janzen, who is the GNP's unsalaried founder,
designer, biologist, arm twister and grunt. He is determined not only to
conserve bits of surviving tropical forest but to restore vast areas of it that
disappeared long ago.

Unlaced boot tops flapping, he leaps down from a boulder and points
to a foot-high tree seedling he recently planted amid the pit's scrubby
vegetation. "This site is going to prove you can grow trees anywhere," he
says. After an hour in Janzen's company, I am inclined to believe him.

Dr. Daniel H. Janzen, who is a professor of biology at the University
of Pennsylvania, has spent his life studying the relationships between
tropical plants and animals. In 1984 his unique blend of theoretical and
field biology—part Charles Darwin, part Daniel Boone—won him the
Crafoord Prize, the Nobel of natural science. This year millions of
viewers were intrigued by a BBC television film about his world, shown
in the United States as part of the PBS series *Nature,* and visitors like
Prince Philip have begun to drop by his rented tin-roof shack in the
tropical backwoods.

Janzen may be a new superstar of international science, but he doesn't
go in for the celebrity lifestyle. Much of his $100,000 Crafoord award
went to buy land for the GNP. The rest paid for the installation of a
phone line and electricity to allow him to run a refrigerator and some
computers at his ascetic headquarters. When I arrived there this morning, Janzen, fit and feisty at forty-nine, bounded out of his house to
shake hands, startling the capuchin monkeys nattering in the treetops
and a pair of iguanas displaying themselves by the porch. He wanted to
know if I had brought my own lunch. He said he would be very busy, but
I could tag along if I stayed out of his way. Tedious bourgeois pleasantries
dispensed with, we climbed into his Landcruiser and set off to inspect
trees.

Now, many trees later, I say good night. I have heard a rumor that
tomorrow Janzen will be taking some visitors to a wild beach at the
GNP's Pacific boundary. Thinking of the winter slush I had left back

home in the States, I ask what time we will be leaving for this tropical paradise. Janzen's eyes flash blue lightning. "The people who are going to the beach are vacationing," he says, enunciating each word so that even a feckless journalist can understand. "I am not on vacation. I am going to work tomorrow."

Janzen probably picked up his social graces when he was an MP in the Army. He relates to *Homo sapiens* best when teaching, and he has even won awards for it (he spends one semester a year at the university). When discussing flora and fauna, he is animated and engaging, and a good question delights him. But when the conversation shifts from his interests, he becomes bored and remote. He often ignores personal or peripheral remarks, perhaps because, as he says, "even when I'm talking to you, I'm thinking about something else."

Janzen has always been happiest alone in the woods, first as a solitary child, then as a hunter and trapper, eventually as an ecologist. He hates to interrupt his long-term researches on mice, moths and seeds and has them planned up to the estimated time of his death in thirty years. Until recently he had pursued these studies for sixteen hours a day, seven days a week, in the Costa Rican forest. But these days Dr. J. is willing to shake hands, kiss babies and even give interviews in behalf of the tropical forest he is growing from scratch. While more sensitive souls despair over preserving what's left of the tropics, ornery Dan Janzen slogs away, resurrecting massacred natural environments.

The next day I arrive at Janzen's house just in time to follow him on one of his daily rounds of his backyard forest: Santa Rosa National Park, the thirty-nine-square-mile hub of the much larger GNP. For almost three decades, Janzen has practiced this sort of "muddy boots" biology, assembling his observations of minute details of plant and animal life into an ecological picture of the tropical forest.

To get the quality and quantity of information he needs, Janzen uses all his senses to pick up things others don't even notice. Pointing to some feathers blended into the leafy ground cover, he says, "That's where one endangered species, the ocelot, ate another, the curassow—a big bird something like a turkey." Mashing what looks like a charred coffee bean between his fingers, he says, "Caterpillar shit. If I wanted the guy who did this, I'd hang out right under this tree." He spots, then proffers, an unattractive brown fruit bristling with seeds; despite its fecal look, it's delicious, like a raisin crossed with a raspberry. Today's jaunt is proceeding tranquilly enough, but that's not always the case. On a previous trip, Janzen came upon a five-foot boa constrictor lunching on a fawn; he yanked the young deer from the snake, weighed it and put it back.

This section of Janzen's forest, which has been preserved in pretty much the same condition it was in 400 years ago, feels unexpectedly familiar at first. Except for a bright yellow monkey shrieking in the canopy, it looks more like California than Green Hell. *Tropical forest* conjures up visions of rain forests, but those wet evergreen environments were once paired with dry deciduous forests, like this part of Santa Rosa, in which the trees shed their leaves for part of the year. Dry tropical forests covered much of Central America when the Spanish arrived in the sixteenth century. Because it was better suited than the rain forest to farming and civilization, the settlers slashed and burned the dry forest to two percent of its original area, while more of the inhospitable rain forest remained intact.

"Yes, the tropics are damaged, but it's not all doom and gloom," says Janzen, tramping along. "Large parts can be restored. Yes, dry tropical forest has been mostly removed, but it's tougher, so it's very regenerable."

According to Janzen, regenerating a tropical forest is mostly a matter of protecting lots of land from people. Given time—25 years for a closed canopy, 200 to 300 years for the cathedral look of a primary forest—animals and wind disperse the seeds that take care of the job. The essential first step of buying up all the properties for miles around Santa Rosa is well under way. Despite its Wild West landscape and cowboy culture, ranching in this part of the Guanacaste Province is no longer very profitable, and the local landowners are willing to sell. Once a farm has been ransomed, agricultural fires no longer kill its vegetation, and wild animals, protected, return to help spread seeds.

To nudge nature, Janzen uses rare remnants of existing dry forest as giant nurseries; park personnel sow their seeds in the burned-out pastures. Soon thousands of cattle will help play Johnny Appleseed in the ugly-duckling transition between pasture and woods: They'll eat down the tall grass that spreads fires and prevents small trees from growing, defecate seeds and show the local cattlemen who want to remain in the vicinity of the GNP how to convert their tired ranches into more lucrative and ecologically beneficial forestries.

Playing games with Mother Nature doesn't trouble the pragmatic Janzen. He will consider any scheme that can help turn his patchwork of farms and woods into Central America's largest dry tropical forest, linked to the rain forests on its volcanic edges. "The whole world is already *unnatural*, if that word means 'perturbed by man,'" he says. "The philosophical issue now is what kind of unnatural do we want?"

The following morning the biologist Winnie Hallwachs, Janzen's colleague and consort of almost ten years, is doing some baking. She says that since the death of their unhousebroken, dog-size pet rodent, "the house is pretty spiffed up." It is strangely cozy—a mixture of bohemian flat, garage and zoo that smells of mice and brownies. It's webbed with lines hung with drying towels and plastic bags filled with household goods and larvae. Skulls, *New Yorker* cartoons, odd bones, posters, feathers, moths and wire cages cover all surfaces. "What's that?" I ask, pointing to scorpions and snakes floating in murky mayo jars beside the stove. "That's science," says Hallwachs firmly.

Hallwachs, thirty-two, could have been sent by central casting to play a sensitive, soft-spoken yet spunky Jane to Janzen's Tarzan. Her Quakerish beauty shines through a bluestocking's baggy clothes, long hair and glasses. Her good nature shows as she talks about her decision to slow down work on her doctoral thesis because "the park needed a woman to make lemonade for visitors and temper the young-silver-back part of Dan." Although she's an unofficial collaborator on the GNP's scientific development, her most obvious—and crucial—contribution is helping Janzen cope with the many humans involved, from cooks to legislators.

The two became acquainted after she took his course at Penn and volunteered to help with his Costa Rican projects. "He said, 'You're welcome if you can pay your own way,'" she says. "If I can't use you as a slave, I'll try to find someone who can.' My initial fascination was with the way his scientist's mind worked—it wasn't personal."

Although she considers Janzen and herself "a perfect couple," Hall-wachs admits that life with Janzen isn't always easy. "People matter less to Dan," she says, "and there's a part of him that doesn't respond to the usual social triggers. He's a difficult person, but I doubt there's been anyone who's made substantial changes in the world who hasn't been difficult."

Janzen emerges from a back room, where he's been working at a computer setup since 6:00 A.M., dons a kind of miner's headpiece fitted with a light and begins his mouse inspection. Three times a year for the last six years, he has caught varying numbers of mice in 529 live traps and recorded their physical condition and location. "They can bite quite hard," he says fondly as one subject protests its checkup. Shaking her head, Hallwachs says, "The photographers who've come here have gotten some beautiful pictures of the mice, but somehow they never end up in the magazines."

Hallwachs and Janzen are immune to the glamour of local monkeys and jaguars, which they call "boring." Their goal of understanding why the forest looks the way it does depends on determining ecological basics summed up as "who eats who and shits what where." To gather that information, they work from the bottom of the food chain up, studying the flora and the humbler, more numerous fauna that complex, endangered animals eat. His favorites are mice—major seed eaters—and caterpillars, major leaf eaters; hers is the agouti, a five-pound rodent something like a guinea pig.

Using ingenious methods, such as locating the agouti's caches of buried food by following thread inserted into hollowed-out fruit pods, Hallwachs has discovered that because the agouti's sharp teeth enable it to chew the big, tough pods, the small mammal reliably disperses seeds that are too big for even the largest Costa Rican animal to swallow, and so determines the location of many of the forest's largest trees. Her enthusiasm suggests that Hallwachs considers the agouti a kindred spirit.

"Oh, yes!" Hallwachs says. "Biologists pick their subjects according

66

to how they themselves live—Dan tends to approach his as a trapper. Agoutis are shy animals, but they're fascinating when looked at closely. Getting a new bit of information about one is like, oh, finding out your best friend is gay—it opens up the world a bit."

Janzen likes mice, but he loves moths. After ten years' work, he, Hallwachs and their helpers are [in 1988] about a third of the way through the identification of Costa Rica's 15,000 to 20,000 species. He points to the bags of larvae and leaves hanging above his head. "We're just waiting to see what turns up," he says. Hallwachs says that a few hours with moths is Janzen's equivalent of a day at the beach; his pleasure in his subjects is evident as he holds up a big insect that looks exactly like a mottled brown leaf. "Look at the way it hangs upside down by that one stemlike leg," he says. "If you were to brush it off a branch in the woods, it would fall like a dying leaf, floating to the ground in arcs, then lie perfectly still for ten minutes to avoid danger."

Janzen insists that he and Hallwachs can pin 2,000 moths in a day. She claims 1,200: "Daniel, I've counted the pins," she says. "I know what we've done." Janzen says, "I'm talking about what we *can* do."

The dauntingly ambitious GNP project demands a can-do attitude. Although it has attracted national and international acclaim and support, the mechanics of manufacturing this tropical forest seem to depend on a handful of Costa Rican ex-farmers, the occasional visiting scientist or volunteer, Hallwachs and Janzen. Hallwachs remembers a night when Janzen "dreamed he was climbing down a cliff, and someone in front of him went out of control and fell. He woke up very disturbed. I wondered if it had something to do with the park." She recalls another of his dreams: "He jumped off a bridge, but he didn't die. He couldn't actually drown."

"No, I don't get discouraged," says Janzen, jumping into the Landcruiser to do some errands. It won't start. "There's no question that GNP will work—it's only a matter of when. My goal is to do what has to be done to make that event occur, not to produce an appearance or depend on finished stages." I help push while he jump-starts. "My father taught me the importance of long-term goals. If it's worth doing, do it right."

The elder Daniel Janzen, a former director of the U.S. Fish and Wildlife Service, brought a Teutonic thoroughness and determination to his conservation career that live on in his son: When nine-year-old Danny started collecting butterflies, his father set him up with entomologists at the University of Minnesota so he would learn to do it right. But Janzen thinks his imagination and ego come from his artistic but emotionally unstable mother, who, he recalls, was prescribed drugs to stabilize her erratic moods.

Early on, Janzen created his own world in nature. He says he has always been aware of being different from others. "For as long as I can remember, I've felt that the natural world was my friend and that people were not my support base," he says. "I grew up without friends because people were not useful or interesting to me—the woods were. I didn't discover people till I hit my twenties. My parents would say, 'Danny, don't you want to take a girl to the movies?' But I had no interest in movies or parties. Or in females—they didn't want to fish and hunt—until I met my first wife when I was twenty. Like me, she was working in a national park."

Whatever first drew Janzen to nature, the outdoors soon provided him with education, recreation, even role models. When he mentions having visited the site of the wilderness scenes in *"Crocodile" Dundee*, I remark that the Aussie version of Boone or Crockett seemed especially popular with men. "That's because he's an alternative to destructive male stereotypes like Rambo," says Janzen. "The hunting and trapping I did as a kid—figuring out where the deer is going to be when and why—was experimental biology. What ecologists do—figure out the relationships between organisms—goes back 10,000 years, to the Pleistocene hunters."

Janzen had no formal training in biology until graduate school at the University of California at Berkeley. His boyhood passion for butterflies had taken him to Mexico at fifteen, and he returned there years later to write his doctoral thesis. "My life is a series of lucky accidents that I recognize and make use of," he says. One provided the subject of his dissertation. While walking down a Mexican road, he saw a beetle alight on an acacia branch; a stinging ant immediately repelled the big bug.

68

When Janzen looked closer, he saw the bush was covered with ants. Later he sprayed a pesticide, and within a few weeks the bush had been denuded by herbivores and was dying. Next he discovered that if the ants were deprived of acacias, they starved. Janzen concluded that in exchange for the acacia's nectar and sheltering thorns, the ants repel the herbivores—a discovery that transformed him from an entomologist into an ecologist.

His tropical researches brought Janzen to Costa Rica in 1963. Since then a good bit of his astounding academic oeuvre has poured forth from his humble headquarters there. "You must not only have good ideas—you must be able to package and sell them," he says. "Lots of scientists die with eighty percent of what they know locked in their heads." There seems little danger of that happening with him: He has already published some 262 papers, from "Why Food Rots" to "No Park Is an Island," and has designed the premier course in tropical biology, taken by students at forty Costa Rican and American schools and universities. His sheer productivity partly explains Janzen's edgy relationship with certain peers; his little way of pinning moth specimens while listening to their lectures doesn't help.

Some scientists object to Janzen's hands-on, intuitive approach to biology. "Study nature, not just books," he says in the governessy tone of an academic lifer. "I just write about what I see, which irritates some biologists because it's not cerebral, not statistical. I go for a walk, look closely and figure out what's going on. Sometimes I just notice things subconsciously."

He brakes the Landcruiser beside a large tree and picks up a dark brown hard-shelled fruit the size of an orange. He is clearly stimulated by objects like this seed and, like the great Victorian naturalists, is a fanatic collector. "Now, anyone could have logicked out that for a tree to evolve that produces fruits as large, hard to eat and oddly placed as this, mega-fauna—giant mammals, like mastodons, giant ground sloths and glyptodonts—capable of chewing them up and dispersing them in dung must have evolved along with them. But I never realized that until I noticed a crescentia seed like this one germinating in some horse shit. Today's horse is doing the job of the mega-fauna!"

What happened to the big beasts? "Pleistocene hunters," he says. "Man has specialized in replacing the wild animals and vegetation he finds with his own. Get them to show you a Clovis point in the American Museum of Natural History—absolutely beautiful stone blades that were no longer made after 9000 B.C. Anything bigger than a cow must have been exterminated by then. Those guys were incredible hunters!"

The mega-fauna theory is vintage Janzen science: personal revelation based on an observation and intuited in an instant in 3-D living color by an ex-predator.

The next afternoon Janzen goes to Cuajiniquil to buy some fish. He is so engrossed in forests that he seems unmoved by the GNP's marine resources, including this pretty little fishing village, eight offshore islands and miles of pristine beaches along the park's Pacific boundary. Cuajiniquil interests him mostly as a human experiment: Its residents are the GNP's educational guinea pigs. Its school's biology teacher, a marine biologist employed by the park, lives in the village; he also instructs the kids in ecology by day and their parents by night.

"Some people say the third world doesn't care about parks, but parts of society in the tropics do want them," says Janzen. "In Costa Rica ten percent of the land is conserved, versus one percent in the U.S. But parks need beaches as well as educational programs. They must be user-friendly if they're to survive because people won't support what they don't understand."

Janzen describes an experience in 1985 that made him realize the importance of conservation education and hospitable parks and that helped give birth to the idea of the GNP in 1986. The Costa Rican government asked his opinion about what should be done with 1,300 gold-mining squatters whose practices were destroying Corcovado National Park. "For the first time, I studied people," Janzen says. "One day I said to a miner, 'Remember what this muddy stream used to be like? Lovely, clear, full of fish and little shrimp?' He thought for a minute, then said, 'There are lots of shrimp in the ocean.' Immediately, I saw that conservation was an intellectual battle." On Janzen's recommendation the government released park personnel from all over the country to educate the miners about why they had to go; in the end only 280 had to be turned out.

Janzen takes a sharp curve into the village. "I got a hundred-pound road-kill tuna on this spot not long ago," he says, cackling. "Fell off a fisherman's truck." Accomplishing several things at once—talking to a reporter while delivering a message while stopping to take pictures and inspect seedlings on the way to buy dinner—has put him in good spirits. I yell prying questions—Does he have any vices?—into his good ear, and he hollers back, "I do buy a lot of film. I don't drink or use drugs—that's like pulling a random wire out of a computer, and I want mine intact. I chased women at one time, but I found that it had no return."

The traits that make Janzen a first-rate biologist in the wilderness—toughness, goal orientation, the ability to tolerate isolation, workaholism—can seem misanthropic in more conventional settings. Aside from his camera, his MacIntosh and Hallwachs's company, his way of life is as austere as a monk's and more than a little unusual. But practices that at first appear merely eccentric have often been calculated to provide him with maximum efficiency: For example, he lives in his office when he is teaching at Penn because, he says, "if you're lucky enough to get paid for what you like to do, why waste a half hour to go home?" Like the happy starving artist or anchorite, he has eliminated all but what he wants to do. "Most people's interests are divided—job, family, recreation and so on—but mine are not," he says. "I'm not concerned with traditional rituals."

Janzen's nontraditional ways seem especially hard on relationships: Two earlier marriages and two children were lost to the rigors of tropical biology. "A male counts his fitness by how many copulations and progeny he has, a female by how many resources she can attract for her progeny," he says. "Most marriage relationships are business contracts in which each spouse sees the other as an employee. The female rents her uterus in exchange for a paycheck."

Segueing from females who interfere with work to Hallwachs, who augments it, Janzen mellows. "I'm able to do so much because of Winnie," he says. "Before, I could only work in bursts—I wasn't psychologically organized and content with the world. A lot of the ideas for the GNP are hers, but she doesn't push that. She's smarter than me, interested in the same things, and she understands the stresses. If we have to stay up all night and pin moths, she'll do it because she feels the same way about goals."

At the home of a fisherman, Janzen buys his dinner in loud, Midwestern-accented Spanish. He seems more congenial among Costa Ricans than with fellow gringos, yet he'll admit to no particular attachment to his generally delightful neighbors or to his second home. "Costa Rica is simply a place where I can do all the things I want to do," he says. Later it strikes me that Janzen's deeds may belie his brusque words. When we get back to Santa Rosa, the park gates he hates and wants removed have been locked for the night. A family of campers, traveling in the heavily laden Costa Rican style, with a full complement of *niños* and grannies, cheers as he unlocks the barrier. The father pumps Janzen's hands, saying, "*¡El doctor!* You are Don Johnson! The man with the snakes around his neck! Your work is very important!" After many felicitations the family heads toward the campgrounds. Janzen, too, is delighted: "I just made twenty friends for the park!"

Before leaving Santa Rosa, I have coffee with Janzen and Hallwachs. We discuss her latest PR efforts, concentrated on smoothing some rough edges off her mate's fund-raising image. The invitation for a recent Manhattan event announced that Janzen would appear in "native dress"—work pants, boots and work shirt—but Hallwachs coaxed him into a jacket and tie. "Dan's always been able to stand out," she says, "but he's not very good in that habitat."

The numbers suggest otherwise. When the GNP was launched in 1986, the project needed $11.8 million to buy the necessary land; today only $3.5 million remains to be raised. Support has come from foundations and other big donors, but small private gifts are also essential. "There's no mystery about preserving and restoring nature," says Janzen. "We've had the technology and will to do it for more than twenty years. We just need the money."

He is confident that well-informed people will support conservation. "Have you ever noticed that every house contains pictures, figurines and so on of flora and fauna?" he asks. "Humans are hard-wired to be sensitive to the natural world—to a stand of trees, moving mammals, red fruits and flowers. The forest is our most complex environment—much more so than New York City. Losing it would be like losing color vision. Just as music and art do, it relieves boredom and provides richness and intellectual stimulation. It makes life worth more."

Janzen's urgency about completing the fund-raising and administrative groundwork for the GNP seems partly motivated by his desire to hand the project over to the Costa Ricans, after which he'll go back to full-time biology. "Dan's on his fifth or sixth life," says Hallwachs. "There was a time when he had a house and two kids and mowed the lawn. Then he was part of the social experimentation of the sixties. He was an independent biologist for a long time. Now he's very involved in a huge public project. Soon the GNP can forget about him, and he can go back to science."

When I ask Janzen what his most rewarding moments as a scientist have been, he mentions a few eureka-type revelations, like the megafauna theory. But he is most enthusiastic about experiencing the annual onset of the rains. "The whole world goes from being dormant to being alive in two days," he says. "Suddenly you experience the sights, sounds, smells of zillions of species. The frog behind your toilet answers a chorus kilometers away." Hallwachs chimes in: "A sheet of iridescent green stars—fireflies—hovers absolutely silently a foot above the ground at night." "Around that lamp," Janzen says, pointing above his head, "would be a mass of termites so solid that you couldn't see the bulb. It's like the whole American spring condensed into forty-eight hours." Nodding, smiling, they speak as one person, linked by shared experiences few others can even imagine.

I wonder aloud if Janzen has ever been motivated by the transcendental, even religious, considerations that draw many to ecology. The idea amuses him. "There's no meaning to life other than trying to make it better for other people," he says. "I figured that out when I was ten. All I've ever wanted was to be really good at what I did." He points at the forest fading in the tropical dusk. "I want that to be there a thousand years from now," he says. "That's enough for me."

Power Play Endangers Hawaii's Rain Forest

Bill McKibben

Consider, as you read this story, two images:

The first is the Hyatt Regency Waikoloa, a hotel that opened in 1988 on the Kona coast of the Big Island of Hawaii. Arriving guests at this $360 million pleasure dome have, in one visitor's words, "all sorts of ways to get their bags to their room—electric boats, electric monorail.

Walking is not an option." The Hyatt Regency Waikoloa has been described as "the most spectacular resort on earth." It also draws four percent of the island's peak-load electricity. One hotel—four percent.

The second is of a gracious dinner out on a breezy porch. As the sun sets, someone flips on a light—a light powered by a twelve-volt solar panel on the roof. The television runs off the same panels; the sun also heats the water, so it's only lukewarm this week. But not bad.

On the Puna side of the big island, the Kilauea volcano slopes down to the ocean. It is the nation's youngest and most active volcano; at the moment it is pouring lava from one vent into the ocean, adding as much as an acre a week to the size of the island. Not far from where the lava pours into the hissing, steaming sea, there is a rain forest—the largest lowland rain forest left in the fifty states.

In some ways the Wao Kele o Puna rain forest resembles the great ones of South America and Indonesia: Dominant trees, in this case *ohias*, shade a dense understory filled with layer upon layer of lush, almost obscene greenery. But there is one reason that this rain forest is unique, and that is that it sits on the side of an active volcano. It does not lie undisturbed for eons; almost continuously, patches of it are inundated by molten lava, which destroys all plant life and leaves behind a layer of black rock.

As a result, the trees don't grow to the same heights that they do in some rain forests. On the other hand, the forest has learned to replace itself quite quickly after a major disaster. "A lot of people look at the trees here and think they're kind of scruffy—they think it couldn't mean much, but it does," says Hampton Carson, a biologist at the University of Hawaii, in Honolulu. "This forest is unique—it's one of the few places on earth where you can tell where life is coming from."

But the location that makes the forest special may also be its undoing. For, with state cooperation, developers have now started drilling through the forest floor to the hot steam below. They plan to produce power from geothermal steam—perhaps as much as 500 megawatts from dozens of wells scattered throughout the forest.

And they are doing it, so they say, in the name of the environment: They will generate electricity from Wao Kele o Puna without burning fossil fuels that give off carbon dioxide and add to the greenhouse effect. A lot of products, large and small, will be sold in the next few years with the label "environmentally friendly." Some of them will be good and solid—and some of them will be shoddy junk.

To figure out which category this project falls into, one could begin at any number of places. The most obvious is probably the biological. The state of Hawaii is aware that the United States is pleading with poor nations around the tropics not to cut down their rain forests, and aware that rain forests have been described as the most precious planetary treasure, and aware that most tourists would probably just as soon think of Hawaii as Eden—so it has taken great pains to say that this is not a "good" rain forest. The governor recently labeled it class C forest; University of Hawaii botanist Charles Lamoureux, who is paid by the developers to do surveys of the land, refers to it as A-2 forest, less pristine than nearby groves of A-1.

"There are a lot of *ohia* trees," says Lamoureux, "but the understory is heavily infested with guava," an exotic that threatens to crowd out other species. This may be true. On the other hand, the state itself declared the land as one of its first chunks of Natural Area Reserve in the 1970s—a designation discarded a decade later when the geothermal drilling was proposed.

And if the forest is so worthless, it's unclear why the state is enforcing what it calls the "strictest" regulations designed to "mitigate" any environmental damage from the drilling. Before any road clearing takes place, for instance, Lamoureux and some assistants go in with the surveyors and try to route them around anything of special significance. "We had them move the drill pad 300 feet to avoid a forest where there were some birds," Lamoureux says. (Unfortunately, when they went to clear the land for that very first site, surely aware that everyone was keeping a close eye on their work, the developers wiped out eight acres instead of the allotted five. They've since closed the drill site to outsiders, including reporters; cops wait near the gates and write down your license plate number if you drive up.)

And the state argues that the developers are only going to clear about 300 acres of 30,000. "It's only about one percent—that's so little," says

Lamoureux. Unfortunately, much of that acreage will be spread out in the form of roads—and these roads, and the men, animals and trucks that traverse them, may well be perfect avenues for precisely the sort of weed species the state claims are currently degrading the forest. "You'll end up with a honeycomb of roads, corridors and steam pipes, and they'll allow the weeds right into the heart of the forest, where the plants aren't used to foreign competition," says Russell Ruderman of the Big Island Rainforest Action Group. It's like arguing that your veins and arteries only take up one percent of your body so it really shouldn't do much damage to eat nine or ten sticks of butter a day.

Class C or A-2 or whatever you want to call it, this is gorgeous, dense forest, undercut with cracks and rifts from the lava flows. I saw it in the company of Michael LaPlante, who lived on the edge of the rain forest until the geothermal plans drove him away. A burly man, he is given to saying things like "that just irks the snot out of me." (He once attended a hearing on the project wearing a shirt he had drenched in hydrogen sulfide, a rank gas that is a byproduct of geothermal drilling. "I found it in a joke store—it was called Morning Breeze Perfume," LaPlante says. "As soon as I sat down, the lady who was supposed to be taking notes for the meeting said, 'I can't do my job under these conditions.' 'Exactly my point,' I said. 'I can't *live* under these conditions.'") Anyway, LaPlante knows his way around this forest. "This is a friendly jungle—no snakes here, nothing to bite your butt," he says. "Look at that *ohia* over there— it probably dates from before Columbus."

There are other ways to look at all this—Palikapu Dedman's way, for instance. A native Hawaiian, he tends to get angry at public hearings and shout. "Our religion *is* a healthy environment," he says. "That's our life, our customs, our medicine." Dedman helped start the Pele Defense Fund named for the volcano god, one of the important deities of the Hawaiian pantheon. I have been places before where it is a little difficult to understand why the native people worshiped a *particular* mountain or hillock or lake—but on Hawaii that is not a problem. As far as the natives who lived there knew, the Hawaiian islands, the most remote spot on the planet, were the entire extent of the earth. On one of the islands, Hawaii, there were volcanoes erupting regularly, and in between eruptions, there were often lakes of fiery lava rippling in the craters and always thousands of vents of steam working their way up to the surface; these eruptions were obviously building up the island as the native people watched. It would have been curious if they had *not* venerated the volcano.

When the Pele Defense Fund tried to raise this point with the courts, however, they were told that they needed a "site-specific" religious practice in order to protect the land from intrusion. "But Pele religion doesn't work that way," says Tom Luebben, one of the PDF counsels. "You may go here one day and there the next, depending on what's happening, where the steam is."

If all this strikes you as too mushy, perhaps some hard economic arguments are in order. A consulting firm hired by the Pele Defense Fund estimates that the total cost of the project, including a cable run at a previously untried depth to the island of Oahu, may run upward of $4 billion. That's between two and three times what the state originally estimated. (The Hawaii Electric Company currently has bids in hand from consortia of giant companies but won't reveal the dollar totals.) "When you make a commitment to costly, risky technology, you're also making a commitment to pay it off," says Robert McKusick, a natural-resources consultant.

There are also technical questions about how long the steam will last. One key scientist at the Hawaiian Volcano Observatory says it may be quickly depleted, forcing the developers to pay for their expensive undersea cable by continually expanding their drilling operations into new corners of the forest. Some doctors insist that the hydrogen sulfide that is vented from the wells during initial drilling—and in the event of an accident—will do great damage: Two or three times the neighborhood around a test plant has been evacuated due to hydrogen sulfide emissions.

State and company officials answer these sorts of charges by stressing that they have new technology that will control emissions and that other parts of the world have sustained geothermal drilling for decades. But their argument for the geothermal development eventually comes back to a single point. "We don't find any other alternative source of electricity large enough to put a dent in Hawaii's consumption," says Roger Ulveling, the state's director of business and economic development. Hawaii depends heavily on imported oil for electricity generation. Its leaders say they would like to lessen that dependence. And so they want to mine one of their most plentiful resources—the heat of the volcanoes.

At first blush, this is a compelling argument. Oil is not such sweet stuff—quite aside from the Joseph Hazelwood factor, oil contributes huge clouds of carbon dioxide to the atmosphere, which may well be raising its temperature. If the choice were between oil and geothermal, it would be a much closer call.

But listen to Robert Mowris for a while. He is an energy-efficiency expert from the University of California at Berkeley, who was traveling through Hawaii last month [April 1990] carrying his report on the proposed geothermal complex and a suitcase full of light bulbs. These light bulbs, which he would pull out at the slightest provocation, screw into sockets for incandescent bulbs. But they use electricity like fluorescent bulbs, which is to say, slowly. Mowris also had panels of special glass that would greatly reduce the need for air conditioning in Waikiki's hotels. Mowris had gadget upon gadget, all simple, all available. All together, he said, they could cut the island's energy use forty to sixty percent and at a cost per kilowatt-hour five to seven times less than building the new supply.

Say Mowris is only half right—that's still a lot of power you wouldn't need to cut up a rain forest to generate. Traditionally, though, utilities have made their money in the same way that every other business does—by selling product. Under current rate setups, Hawaii Electric can no more make money selling power to people with thrifty light bulbs than *Rolling Stone* can make money not selling magazines.

Something else worth considering: Hawaii has another asset besides lava, and that's sunshine. If there was ever any place where people could take advantage of solar power to heat their homes, it's Hawaii. And some do. LaPlante has had solar panels on his house for fifteen years. Jim

Albertini, a farmer and activist in the rain-soaked Puna area, runs his spread off the power that hits his roof. Quite a few people live "off the grid"—relying on the sun or the steady trade winds for their electricity.

Even in Hawaii, though, where solar energy should work if it works any place on earth, most people get their power from the same sources as you and I. Ulveling and other officials boast endlessly about the number of solar-heat pumps and water-heating systems in Hawaii (more than most places, which seems natural). But the data published by the Department of Business and Economic Development itself shows that after a surge of conservation following the OPEC embargoes, per capita electricity use has resumed climbing quickly in the islands. And the percentage of housing units with solar water-heating systems—just over ten and a half percent—has not increased [as of 1990] since 1985.

Why is this? The reason, I think, is that having a solar collector on your roof is not quite as "nice" as having a line running into your house from some power plant. The water in your shower may not be quite as hot. You can't run your light bulbs and your appliances as many hours a day if it's cloudy out. It's not as utterly and overwhelmingly convenient. When I was talking with Lamoureux, the botanist who works for the geothermal development, he said something almost as an aside that struck me. "I'm not out beating the drum for geothermal," he said. "Left to my own devices, I'd probably say, 'Leave the forest alone, and turn off your lights.' But that's not going to happen. I could live more uncomfortably than my neighbors, but I'm not sure that's what I want."

Which made me think of something W. S. Merwin had said a day or two before when we were sitting in the semidark of evening in his living room, which is lit in the day by the sun and at night, if light is needed, by the sixteen solar panels on his roof. Merwin, who lives on Maui, actively opposes developing the rain forest. He's also a Pulitzer Prize–winning poet and a man who seems able to sum up the issues of energy, forestry and personal responsibility that lie at the center of this, and most other, environmental arguments.

When he moved to Maui, his land was scrub. On it he cultivated rare species of palms from around the world, some of them extinct in their proper habitats. "This tree is from Isle de Reunion," Merwin says. "This is a Guinean oil palm. This has the largest leaf in the vegetable kingdom, fifteen feet across." It is a daily labor of love. "I've grown up all my life seeing what people do to the earth, and it makes me sick," says Merwin. "This was a chance to see, if I had a piece of land, how I would treat it." With little money and a good set of hand tools, he has built a reasonable set for whoever wants to turn Genesis into a movie. But he recognizes that it is a garden. "Only a forest makes a forest," he says.

"What's the purpose of a rain forest?" Merwin asks. "The first answer is, There is no purpose. Life doesn't exist because it has a purpose, it exists because it exists. But the second answer is, *Any* species is valuable beyond any way we can assess it. When we look at the world around us in a particular way, that's what we are. If we look at it as something to exploit, that's what we become—exploiters. If we look at it the other way, it's a sort of endless relationship."

In Brazil they cut down rain forests for cattle ranching; in Hawaii they "bisect" them so you can ride the electric monorail to your hotel room. The question is, How much forest will be left when we finally come to our senses?

Rain Forest Journal

Tom Hayden

> To read or listen to most accounts of Amazonas is to conclude that only a maniac would ever set foot out of doors.
> Peter Matthiessen, *The Cloud Forest*, 1961

So bring it on.

If you want to get away from the L.A. scene on New Year's, let me suggest Manaus, the 350-year-old port of entry to the Brazilian rain forest.

Personally, I have come here to gain distance from defeat and depression. For the past year I labored to pass the Big Green (California's environmental initiative), only to lose to voter anxieties over economic recession. What to do? I believe the environment has to become more than just another issue. We are living through a period of the greatest ecocide, the greatest era of species extinction, since the dinosaurs 65 million years ago. Environmental consciousness needs to be more central in our lives, an ethic about which we are passionate. But how?

Maybe the wilderness has an answer. I don't. And it's lonely.

Thankfully, I am traveling with Troy, my seventeen-year-old son; it's a male initiation ritual Robert Bly would appreciate. Even though I live with Troy, I'm shocked that he's so tall now, six foot two, and 165 pounds of muscle. He's a student at Santa Monica High, with a keen curiosity, a sense of humor and a tight gang of friends who tend toward baseball, Nintendo and rebellion, the kind who play "Fight the Power" while going to the prom in their tuxedos. The kid within him expects this trip to be a wild adventure, an Amazon safari, and equipped with camera, he's planning a class report. But I'm becoming aware of the man evolving in him, too, the person who'll be leaving the nest for college this year. We may not be together quite like this again.

Troy is reading Joseph Conrad's *Heart of Darkness*, the classic novel about a white adventurer who becomes mad from tampering too deeply with the truths of the tropical jungle.

This journey also has its formal agenda. There is a Smithsonian-Brazilian research project in the rain forest where I will learn more about the critical need for research on threatened species. And Dr. Noel Brown, the head of the United Nations' environmental program in North America, has encouraged me to visit a cooperative where local people are trying to find American markets for rain forest products like Brazil nuts.

Manaus was built on one of the many fantastic dreams that brought explorers to Amazonia, that of rubber production. Until the English stole seedlings from the rubber tree, replanted them in Malaysia and ruined the Brazilian rubber boom, Manaus was a city where men lit their cigars with dollar bills, where electric trolleys were running before Boston had them and where one of the world's largest opera houses was built in order to replant the culture of Europe.

The dream of European opera has faded, but the population of Manaus has doubled since 1960 as impoverished Brazilians have moved here to work in a free-trade zone, assembling electronics products. Middle-class Brazilians travel here to shop, take a day trip on the Amazon and return home, where they sell the VCRs they have bought in Manaus to pay for their plane tickets.

We slept five hours in a local hotel, long enough for me to imagine a green and yellow parrot with six-foot wings, before the phone rang and the jeeps were ready to leave.

We bounced along a red clay road north toward what has been called the largest experiment in the history of ecology. Its name is at least the most unlikely: the Minimum Critical Size of Ecosystems Project or, alternatively, the Biological Dynamics of Forest Fragments Project. Although a joint project of the Smithsonian and the Brazilian environmental agency, this venture is mostly associated with Dr. Thomas Lovejoy, a biologist who began observing and banding birds in Amazonia as a graduate student twenty-six years ago. After years at the World Wildlife Fund, he now serves as assistant secretary for external affairs of the Smithsonian. Yale educated, accessible, a scientist who can speak plain English and the creator of the *Nature* series on public television, he is the main rain forest guru for everyone from U.S. senators to Hollywood environmentalists. His creative solution to the rain forest crisis, already adopted by several nations, is exchanges of debt for nature, in which debtor countries, instead of destroying forests to pay international banks, qualify for debt reduction by setting forest lands aside.

The Critical Size project studies isolated patches of forest for the impact of logging or slash-and-burn practices. The question is, how much room does an insect, a jaguar or a creeping vine need to survive and prosper? The answer, Lovejoy's researchers have found, is far more than we realize. A simple colony of army ants needs seventy-five acres; a viable jaguar family may require as much as 1,900 square miles of running room. With tropical forests being destroyed at a minimum rate of fifty-seven acres per minute, Lovejoy's message carries urgency.

We careened along toward Camp 41, eighty-one kilometers north toward Venezuela, then turned left into the forest on an eroded, soggy, boulder-covered, one-lane road until, after four hours, the drivers suddenly stopped, and we continued on foot.

Everything was a green blur—quiet, shadowy, moist. A shriek from somewhere above. Thick vines spiraling up 100 feet toward sunlight.

Palms that alone would fill most living rooms. I tried to focus my eyes and feet on the path, while Troy started looking for the giant snakes— bushmasters—that either kill you or provide the source of great stories.

Accounts of the Amazon had forewarned me of this "apparent emptiness," the "green stillness" of camouflage concealing its life, but I was still losing my balance when after five minutes we came to a tiny clearing.

I recognized Tom Lovejoy, who was smiling and walking toward us. And with him, here in the most remote place I had ever been, was Tom Brokaw.

At Robert Redford's Sundance Institute one year before, Tom and his wife, Meredith, had pledged to visit and film the Amazon rain forest. And true to their word, here they were, returned from a morning of bird-watching. (Another big question: How much do North American songbirds depend on wintering in tropical rain forests?)

We will spend two days here, "touring" the forest. I am fascinated by the guides—two British graduate students affiliated with Brazil's National Institute for Amazon Research. One of them, Andrew Whittaker, is a self-educated, talking encyclopedia of bird lore. He seems to hop through the forest, while the other guide, Nigel Sizer, glides invisibly. Nigel is slender, with large red sores that suggest AIDS or leprosy but come from leishmaniasis, which erupts about a month after one is bitten by an infected sand fly. He shrugs them off, the inevitable wounds of his quest to understand this forest.

Nigel speaks softly of environmental apocalypse. "The dinosaurs had time to evolve into birds," he remarks. "But we don't have the time." It's the 200th anniversary of Mozart's death, he says cryptically. Nigel has been here two years.

We spent the afternoon in the forest, and I still could not adjust to the orgy of life. In an area of four square miles, there are 1,500 varieties of flowering plants and 750 species of trees; by contrast, all of North America has 400 tree species. Meredith and I watched a blue butterfly with giant wings. A pack of capuchin monkeys leaped after fruit growing in the green canopy above us. Troy photographed a column of leaf-cutter ants that have more muscle, relatively, than we do. He's still an urban child, but I detect a growing fascination with this place. Meanwhile, I stared at a strangler fig, which is born in bird shit and somehow grows to curl around a tree until absorbing and destroying it.

Later we came across a seven-foot boa constrictor. Andrew delightedly grasped it behind the head and draped it over my shoulders. Brokaw looked away for a second, and the snake tried to attack him at the waist. With the snake menacing, Brokaw filmed a New Year's greeting to David Letterman.

I wondered if the bats we saw at dusk were vampires, if the mosquitoes carried yellow fever, malaria or elephantiasis and if Santa Monica doctors could cure leishmaniasis. At first, Troy wouldn't even cool off in the camp-side stream because he had read about the candiru, a wormlike catfish that slips into bodily crevices and stabs its pointed fin upward.

All Amazon travelers have chronicled their phobias. But it's not what the forest might do to us that is so troubling. It's what we have done to it.

I have come here, trying to feel the destruction of the rain forest, because intellectually I cannot grasp it. The effort reminds me of trying to comprehend nuclear war when I was very young. The mind gives up, the emotions numb when the horror is too great and abstract. I have resisted thinking in apocalyptic terms since the sixties. But I can't anymore. By candlelight tonight, I once again took out my notes on what is known of rain forest destruction.

At least fifty percent of all living things—some say more—exist in rain forests, the gardens of Eden where our amphibian ancestors perhaps first crawled upon land. There might be 30 million species, but we are only guessing because less than ten percent of such life, the genetic base of all we are, has been thoroughly identified. At least five species, plant or animal, become extinct every day. According to Lovejoy, we are destroying species before we discover them.

You might ask, being self-centered, why we should care about a holocaust descending on obscure species if it will guarantee teakwood floors from logged forests and cheaper hamburger meat from the cattle ranches that replace them.

The answer is that nature is like money: You can live on the interest but have to be careful with the principal. If we destroy our genetic bank, we destroy the diversity of nature, lose countless resources we depend on, risk the integrity of what took evolution billions of years to create.

Let's be specific. Burning the Amazon rain forest in 1987 alone sent 3 billion tons of carbon dioxide into the atmosphere, nearly one-fifth of the total global emissions of this "greenhouse" gas. Deforestation causes equally scary damage to topsoil, removing one-third of that available worldwide for subsistence farmers. This, of course, means more hunger and floods in places like Bangladesh; according to one scientist, up to 1 billion people may starve to death in the next three decades.

Cutting down the rain forest creates a health epidemic in more ways than one. Mosquitoes, which carry malaria, live high in the forest canopy. But when the forest is replaced by roads, construction and swampy puddles, the little malaria carriers descend to our level. Iron-ically, the cure for malaria was discovered in the rain forest 300 years ago when quinine was developed from the cinchona tree by the Amazon Indians. (Oliver Cromwell died of malaria because he thought the cinchona remedy was a Catholic plot initiated by Jesuit conspirators in the Amazon.) The modern world eventually turned to synthetic malaria drugs, a chemical fix, against which the mosquitoes have become ever more resistant. Now malaria is returning with a murderous vengeance, killing at least 2 million to 4 million people around the world yearly. In one Amazon state we will be visiting, Rondonia, twenty percent of the population has the disease.

When I raised these matters tonight, Meredith Brokaw said the health issue was the way to reach most Americans about the repercus-sions of this distant biological holocaust. The story of quinine is undis-puted. But few realize that almost one out of four medicines in our pharmacies literally have roots in the rain forests.

Lovejoy often cites Capoten, marketed by Squibb, which is the pre-ferred prescription drug for regulating blood pressure. It was created as a result of studying how the bushmaster's venom was able to drop its victim's blood pressure. The Eli Lilly drugs vincristine and vinblastine, effective in causing remission of children's leukemia, come from the minuscule, rosy periwinkle plant. It's been estimated that there are at least 1,400 forest plants and insects with cancer-fighting potential.

If you ask what the rain forest has done for us lately, the answer is, plenty, a formidable list not only of "green" medicines but of foods, rubber, oils, fuels of the future and so on. The economic value is hard to quantify but clearly amounts to billions of dollars yearly. And unlike resources like fossil fuels, these are all renewable.

But a scientist like Lovejoy is inspired by more than pragmatic arguments. I am interested in the deeper emotions and motives that drive someone like him. "The rain forest has become more than the rain forest—it's awakening us to the environment," Lovejoy says. "We're disrupting the planetary ecology in ways we can't even understand.

"It's really been about two and a half years since I started getting truly upset," Lovejoy continues, "when the global-warming data was coming in, when you could no longer think of it except as a truly major problem. I think we need a real revolution."

As a biologist, Lovejoy comprehends more than I the importance of the forest as a genetic laboratory; at least fifty percent of the gene pool of the whole world is here in Amazonia. He points out that one mouse chromosome contains as much information as is stored in all the editions of the *Encyclopedia Britannica* ever printed. We would be outraged if someone tried to burn books in the Library of Congress, but Amazon fires routinely burn species to extinction. The less diversity of species, he notes calmly, the greater the threat to our species. For example, if the peregrine falcon had not been around to die of DDT poisoning, the pesticide would have imperiled us far more than it did. "When we overshoot," Lovejoy says, "there will be all kinds of natural disasters and diseases."

Nigel was treating his leishmaniasis with an ointment as we talked by candlelight. "The problem is short-term thinking," he said. "We cut down the forests for so-called development today but threaten all life tomorrow." He muttered again about its being Mozart's anniversary, and I realized it was because Nigel believes that life should be ordered like a

Mozart symphony, a perfect reflection of the universe, and not according to the interests of politicians and generals.

January 3rd

NIGEL IS STUDYING the "edge effect," or what happens to the forest at the boundary of a cut by cattle ranchers or loggers. We drove to an area where a 10,500-acre ranch had been slashed out of the forest in 1980. Like many such ranches, it was abandoned because the denuded soil was of poor quality. It was a financial harvest the ranchers were after anyway, because until recently the government policy has been to provide massive tax breaks for developers in the Amazon.

Nigel analyzes the impact of deforestation on different species. The only increase in numbers, he says, has been among certain butterflies that love light; the number of birds and frogs has dropped. He collects leaf litter from seventy-two traps, beginning at the edge and moving into the remaining forest. There has been a lessening in the natural process of nutrient cycling, he says. More leaves have dropped at the edge, where they are exposed to sun, wind and rain, than deeper within the forest. "A severe implication," Nigel says sadly. "The edge effect is happening. This means the size of a forest reserve or park is even smaller than it looks because the natural process of the forest has been altered. You have to set far more aside than you think, and we don't even know how much more. Here, in one hectare [2.47 acres] there might be 160 species of trees more than ten centimeters in diameter. Where I'm from, in the whole United Kingdom, ten species will get that big."

I am trying to think in the margins like Nigel and Lovejoy do, where a seemingly minuscule difference can loom catastrophic, the way a barely detectable cancer cell wreaks havoc. For example, in Brokaw's interview with him, Lovejoy said: "If you cut and burned all the tropical forests in the world, it would increase carbon dioxide in the atmosphere by fifty percent. There's only a five-degree-centigrade difference between a glacial and a nonglacial period, and it's already a half-degree warmer than a century ago." This is a cryptic way of saying we're helping to melt the polar ice caps by burning the Amazon.

January 4th

I ASKED LOVEJOY this morning why governments and universities don't make a more urgent effort to take an inventory of all living things before they are extinct.

A light of interest went on in his eyes. "A global project," he said. "An idea for the UN Conference [on the Environment and Development, to be held in Brazil in 1992]. We can do it."

The effort, Lovejoy said, would begin with known species. In 1990, for example, 100 experts on flora and fauna of the Amazon met to chart what the conservation priorities should be. Next, he said, "you would look at a map for places no one has studied, reveal the diversity no one has cataloged, fan out over the face of the world with zoologists, botanists, graduate students, and identify the areas where species will die off if there's development. The most important thing is that it would create indigenous environmentalists out of local biologists." The cost, by one estimate, would be less than we spend monthly in the United States to shelter stray dogs.

I am sure Lovejoy will turn the idea into a proposal at the 1992 Brazil meeting: an international brigade of Andrews and Nigels, bringing species loss to world attention. But is there time to *save* these species or only to research their extinction? The UN conference will test whether the world's governments can effectively reverse global warming, forest destruction and ocean pollution, the ills that make the planet a dying organism. I feel a worried hopefulness. "The next ten or twenty years will tell," Lovejoy said, raising his hands in doubt. "There is no greater environmental problem."

To be here is to experience how life began and how it might end.

On the River—January 5th

THE BROKAWS AND Lovejoy left this morning. The forest camp is behind us, and we finally are on the Amazon. Actually, we are at the "meeting of the waters," where the creamy chocolate Amazon merges with the dark cocoa color of the Rio Negro, its largest tributary. Our guide, Cynthia, says the native people believe there are monster fish here, but it's okay to swim. We decline.

This surging river accounts for one-quarter of the world's fresh water flowing to sea. Around us are at least 3,000 species of fish. We will journey up the Rio Negro for three days. The destination is a vast, flooded forest. In a symbiotic, life-giving transfusion, the river fertilizes the forest with silt carried from the Andes, and the forest drops its fruit and seedlings to the fish below. If the forest disappears, of course, most of the commercial fish of Brazil will die with it.

The journey is a lazy one, and as we leave Manaus, the monotony of the forest resumes. A plane or two from the modern world passes over. At river's edge we begin to see isolated thatch-roof houses on stilts; they belong to *caboolos*, poor people of Indian descent.

I see a tiny church and think of the film *The Mission*. Brokaw admires the Jesuits, particularly a priest he met working with gangs in East L.A. How far away that seems, but it's all one path—the ancestors of today's Indian tribes migrating from the tropical forests of Asia across the Bering Strait 10,000 years ago, some settling in Mexico (and later in California) as others moved across the isthmus to this tropical river basin stretching more than twice the size of India, the Indians surviving the environment somehow but dying in massive numbers under European conquests, their numbers plunging from between 6 million and 9 million in 1500 to 1 million in 1900 to less than 200,000 today. Yet everywhere the priests are patiently preaching conversion. But in the end, conversion to what? If liberation theology has risen to defend the poor, could there be an environmental theology that comes to the defense of nature?

Troy is awake. He was expecting something out of the movie *The*

Mosquito Coast and likes the fact that our boat has a cabin. Jokingly, he spreads his arms and declares: "I don't know what these environmentalists are worried about. Look at this space! People ought to be able to live along the river with plenty of room for plants and animals. What's the problem?"

His eyes can tell virgin forest from second-growth now, and he's beginning to confess a liking for the manioc we're eating, the common meal of the local people. You see *caboolos* growing it everywhere. They peel, soak, grind and squeeze out the juice, which contains hydrocyanic acid, and in a few days the pulp is ready to eat. Troy and I are trying to figure out who in the course of evolution learned to remove the cyanide.

We see a house that looks as if it's made of mahogany, with a nice lawn, like a place in Santa Monica. "If there's any drug dealers here," Troy says, "they live in that house."

"Missionaries often have nice houses," Cynthia says.

At dusk we pass a tiny community with lights. Women are washing plates in the river. There is a house with a television, another with a Christmas tree, and a school. "I guarantee the white man is behind this," says Troy. "In twenty years it's going to be Manaus here." As Nigel is into plants and I am into environmental visions, Troy says he's "into the illegal power behind everything here." And in fact the vast, shadowy Amazon is the scene of illicit traffic in most things contraband.

We are at the islands where we will tie into the trees for the night, about eighty kilometers northwest of Manaus. After eating, we venture out in a smaller boat, looking for such nocturnal creatures as alligators. We move quietly into an everglade dark enough to give rise to the most primitive fear. Frogs and monkeys yak loudly. Far away, a dog barks at the unknown. The shadowy limbs just above us could contain bushmasters or perhaps a resting jaguar. Suddenly, our flashlights rivet on the glowing red eyes of a two-foot alligator. We lift him easily into the boat for close examination. He is an innocent child of a threatened species, says our guide, not trained in deceiving the hunters who would happily turn him into a pair of boots.

Rio Negro—January 6th

WE TRAVELED STILL farther up the Rio Negro to a wild archipelago of countless forested islands; I couldn't be sure where the banks of the river were. The river supports 800 species of fish, 200 more than are found in the United States.

The most famous of these is the wicked piranha, which Troy wanted to catch for a cheap thrill. He and I took a small boat and, with hand-held lines and bamboo poles, hooked up bloody scraps of meat and gradually caught a bucketful. No species gives the Amazon its forbidding image more than this eight-inch fish, which looks like a bluegill with fangs.

Our guide for this leg of our trip is a good-natured man named Nonato, who belittled the danger and urged us to swim. We refused. I caught a "vegetarian" piranha, one that feasts on fruit droppings from the forest above. But the rest of our catch was carnivorous, and one bit off the flesh on the end of Nonato's little finger. Whatever was rotting in its mouth infected the bleeding finger. Hours later, Nonato said his finger-tip felt like someone was pounding it with a sledgehammer.

Later that night, Troy started singing the theme song of *Happy Days* while beating the bed with both fists. "I need a little civilization to balance this," he cracked. Later, while reading *Heart of Darkness*, he looked up and said, "Maybe this is the civilized world, and we live in an uncivilized society."

Rio Negro—January 7th

THE WORLD'S RAIN forests are being destroyed by lumber companies, cattle ranchers and governments that dream of selling teak to Japan and hamburgers to America and Europe or paying off debts to the world's biggest banks. Brazil's debt is $113 billion. Since most desirable land in a developing country is owned by a wealthy one percent of the population, a government like Brazil's historically channels its country's exploding population toward the unsettled forests. The result, writes author Norman Myers, is that "the number one factor in disruption and destruction of tropical forests is the small-scale farmer," an "unwitting instrument," just like a soldier in war. Compared with the poverty and unemployment found in the cities, a plot of land in the jungle is appealing to farmers like Odair Lopes Ferriera.

We met Lopes Ferriera by pulling up to his floating dock, about two hours from Manaus, and introducing ourselves as curious North Americans. His view was breathtaking—"the sort of place you can put a $2 million home someday," Troy observed—but for now it was a tiny homestead perhaps like those that dotted our own frontier 200 years ago.

Lopes Ferriera, who welcomed our chance arrival as if he'd been waiting for us, was a short, powerfully built man of perhaps forty years, wearing only ripped shorts on his deeply brown body. He had obtained his plot through a government resettlement program, had constructed two buildings and had hacked out ten acres or so of forest for his crops.

He and his wife like subsistence farming much better than life in Manaus, where, he said with anger, "the rich billionaires get richer. It's going to stay that way, and I'm tired of fighting it." Here, he simply chopped down some rain forest, burned it and planted his manioc, cashews, pineapples and vegetables. He felt free to chop his way to the next river. And if more and more people moved from Manaus, he planned to open a little store on the riverbank.

"We have a lot of problems with environmentalists," Lopes Ferriera complained. "I don't agree with them. I am using the land productively. The real problems are the big cattle ranchers and developers. It's hard work, but I'd rather work hard on this land myself than in a factory in Manaus. This way I can be self-sufficient and live and work for myself alone."

We parted after ninety minutes, Lopes Ferriera waving cheerfully as we headed down the river to Manaus. Perhaps, I thought, the poverty of the soil and the hardship of forest clearing will limit his expansion, but then what? Will he find a way to use and sell products from the forest—Brazil nuts, fruit, vines for baskets and brooms in Manaus—instead of cutting it down? Will he just keep cutting and moving on until the forest is gone and the soil is dead?

Rio Branco—January 9th

> And outside, the silent wilderness surrounding this cleared speck
> of the earth struck me as something great and invincible, like evil
> or truth, waiting patiently for the passing away of this fantastic
> invasion. Joseph Conrad, *Heart of Darkness*

WE ARE FLYING three hours westward to Rio Branco, crossing over what my contacts have called the "road that killed Chico Mendes." (Mendes was the organizer of a rubber-tappers union that made common cause with environmentalists trying to protect the forests; he was shot and killed by local cattle ranchers in 1988. Between 1985 and 1989, there were 539 assassinations due to Amazon land conflicts. Two of Mendes's killers were convicted in a Rio Branco courtroom in late 1990.) BR-364, as the highway is known, was hacked through the jungle to bring the landless to Amazonia, and between 1970 and 1985 they came, the state growing in population from 100,000 to 730,000. With them came cattle ranchers, deforestation and the 1987 fires whose smoke, in NASA satellite photos, seemed to cover the earth.

Such destructive policies are being slowed, I am told. Mendes's struggle influenced the development banks to reconsider the backing of reckless road-construction projects, and the Brazilian government has cut tax subsidies to cattle ranchers and placed the renowned environmentalist José Lutzenberger in charge of the country's ecological policies.

Mendes's key idea was to create "extractive reserves" in Amazonia, where rubber tappers, subsistence farmers and Indians could live by harvesting products from the forest for commercial sale instead of burning it down, an idea the new Brazilian government now endorses.

From Rio Branco, a helicopter will take us to Mapia, at the center of a 1-million-acre reserve. The Mapia community has a $200,000 loan from the Interamerican Development Bank for Brazil-nut extracting, is seeking a joint venture with Goodyear for rubber and is being watched by the UN as an experiment in balancing economic growth with rain forest protection. Mapia's leader is a man whose formal signature is Paulo Roberto Silva e Souza but whom everyone calls Paulo Roberto.

The wilderness around Mapia stretches north and west to Bolivia and, at the base of the Andes, Peru. It is where the Incas fled the Spanish 500 years ago and where, in the sixties, Che Guevara launched his guerrilla campaign for continental liberation.

Leaving Rio Branco, we fly low over the stunning canopy of the untouched forest until, after an hour, we drop into a clearing of perhaps 300 acres. We land in a pasture surrounded by a few dozen cabins, administrative buildings and a large circular structure that seems to be a church. It has a large cross on top but with an upturned crescent and two horizontal bars. Staring at us are the curious, friendly faces of some fifty people, mostly kids, a dog angry at the helicopter and several very spooked head of cattle. I'm not ready for this abrupt entry, but the alternative is to spend four days in a canoe plagued by flies whose bite can give you river blindness.

I focus on a slender man with friendly, riveting eyes, a slight mustache, a baseball cap and a rain forest T-shirt, knowing instinctively he is Paulo Roberto. He steps forward, grips my hand at length and introduces us to a second Paulo, an earnest, bespectacled administrator of the project. Circled by friendly stares, we are taken to Paulo Roberto's house a few yards away.

On the outside, it looks like an unfinished, unpainted two-story barn. Inside, it has the open feel of a communal dwelling in northern California or Vermont. There is a spacious living room, a dining area and kitchen, and a staircase that leads to bedrooms upstairs.

On the wall are large portraits of Saint Sebastian, the Virgin Mary and a black woman walking on water (I am told she is an African goddess of the river). There are smaller pictures of Christ and a book on Amazonia by Jacques Cousteau. There are two photos of an elderly man with a flowing beard. There are no electric lights, and the windows are simply open apertures that can be closed by sliding panels.

Paulo Roberto's wife, Nonata, and several friends are waiting with water and coffee. Physically, Nonata reminds me of Nigel; her very languid body seemed to fit right in with the forest environment. Her hair is long and as black as her eyes. Later, I learn she suffers continually from malaria.

Paulo Roberto and Paulo the administrator, I realize, are too middle-class to be rubber tappers, and I ask about their background. Paulo Roberto, now forty-one, was a Rio de Janeiro therapist interested in "bio-energy," the removal of emotional blockages, who began a "spiritual quest" that took him to Rio Branco a decade ago [1981]. There he met Nonata, joined this fledgling community and moved to the jungle. The other Paulo was originally Paulo Roberto's patient in Rio, where he now stays to write grants, administer the project's budget and deal with government agencies and media.

"Our goal is to be a model by 1992," says Paulo Roberto, referring to the UN conference. The problem of Amazonia, he says, is that forest dwellers have faced a "false dilemma" between rain forest preservation and economic development.

"When we got here in 1982," Paulo Roberto says, "we saw the paradox that we lived in the richest region in the world yet we were the poorest people. Deforestation was coming at the rate of one football field per minute. And we came to a solution on three levels: ecological, social and economic."

By next year [1992], he says, they will have the Brazil-nut production fully operating; at present they can export sixty-eight tons yearly ("Not much," he says, "but we'll reach 200 tons"). Brazil's president, Fernando Collor, will attend the ribbon-cutting ceremony for the co-op's school for 100 children this year. The talks with Goodyear over a rubber contract are promising, and the rubber tappers are coming up with new ways to reduce the costs of extraction. There are forty more products the community can develop, including eighteen types of vegetable oils, medicinal plants, fruits, seeds, straw, soaps, buttons. And they are cooperating with Indian tribes to the north and south: "We can process their nuts," Paulo Roberto says, "and do research on medicinal plants—they know 1,200 medicinal plants."

I ask how he chose this site. Paulo Roberto points to the portrait of the man with the flowing beard and smiles. "He did, that man. He was our spiritual leader, Padrino Sebastiano. He died last year. I am his son-in-law."

Mapia—January 10th

> "Tell me, pray," said I, "who is this Mr. Kurtz?"
> "The chief of the Inner Station."

Joseph Conrad, Heart of Darkness

WE SPENT YESTERDAY afternoon walking around the settlement. The food is good: meat, fish, manioc, vegetables and rice, all produced here. The buildings are constructed out of twenty-seven varieties of trees. The typical dwelling is a spacious hut with a thatch roof, usually out in the forest where the tappers work all week, coming into the community only on the weekend. Troy and I now enjoy the verdant flora that we wouldn't have noticed a week ago: Here, in a few square feet, is a cashew tree, an avocado tree, a coffee plant and something called an inhame that has giant green leaves and roots—the extract of which is good for the blood. There is an urucum tree whose red flowers yield spice, skin oil and facial paint for Indian ceremonies.

"Before this," Troy told me, "nature was just there, something out the window. I never saw its power." He finally made himself jump in the cold, deep-running creek nearby to cool off. We washed ourselves with soap from the copaiba tree while standing on a log in the current. Troy exercised his upper back by swimming against the current until, exhausted, he floated back to the log. "I still don't trust jumping in here," he admitted. "It's not like falling back into a friend's arms."

Afterward we walked on, coming to a low-frame building next to five earthen vats, perhaps six feet deep, hollowed in the ground. "That's where we prepare the daime for our rituals," Paulo Roberto stated matter-of-factly. Daime? Rituals? He explained that daime is the community's "plant of wisdom" and a drink, created from the yagé vine and the rainha leaf, that has been used since before the time of the Incas for "self-revelation." The ingredients are pounded and flattened, then boiled in water in the vats. "There will be a ceremony tomorrow you can attend," Paulo Roberto said, "and we will discuss it again later."

"Is this some kind of weird drug cult?" Troy whispered to me. I felt unprepared for the new information but shrugged and said: "Who knows? I doubt it. They just believe in sacraments and ceremonies, like the Indians." I felt more curiosity than worry. But then again, a voice in my head warned that most crazy cults look normal until it's too late.

In the evening we sat on Paulo Roberto's floor for after-dinner conversation among flickering candles that created an atmosphere of expectation. I decided that Paulo, when fully animated, looked like a bird with jointed wings, short beak, incisor eyes and a wide pelican mouth. He told some scary stories about a local man attacked by a jaguar ("The most frightening is the smell of its breath on you") and the time when anacondas surfaced in the stream where we had been swimming ("You know when they come because the water is shaking everywhere, and you can tell there is death in it"). Troy was spellbound.

During one of the long silences that punctuate conversations in the dark forest, I quietly asked Paulo Roberto if he would return to his earlier comments about a spiritual quest and the daime ritual. I knew that he was waiting to do so and that this was the heart of whatever Mapia is.

"The *seringueiros* [rubber tappers] were, how would you say, compromised as human beings by the rubber barons," Paulo began to explain. "They were treated as nothing. So we decided to found the community on a spiritual basis. Padrino Sebastiano"—he looked at his father-in-law's picture—"drew spiritual knowledge, and self-knowledge, from a very old tradition in the Amazon.

"When the Spanish conquerors came," he continued, "one Inca prince, Atahualpa, got on his knees, offered them gold and was killed. The other, Huascar, went to Machu Picchu [the Incan retreat atop the Peruvian Andes, 11,000 feet above sea level], then came here by the Purus River. He made a big impression on Indian tribes.

"Then the rubber tappers came here in the last century, by the river from the northeast, bringing with them faith in a cosmos, and their own rituals of self-knowledge," Paulo Roberto said. "One of them, Raimundo Irineu Serra, met forest people who knew the ritual, and created our church, whose teachings he passed on to Padrino Sebastiano.

"We need an identity to replace slavery," he continued, "a freedom to know who we are here at this time. The main thing for us is to get this kind of self-knowledge and identity."

I looked up from taking notes, already sensing that my entire journey was coming down to the next question. And, already knowing the answer, I asked, "What is that identity?"

"To be friends of the earth. We have a right to live in harmony with our great Mother."

And how is the identity achieved?

"The big challenge is knowing your inner self and its relationship to your community and to the cosmos," Paulo Roberto said. "For us to live here in harmony with each other and with the environment is a big challenge. It is difficult to describe; it comes from an altered state of consciousness, from the daime."

The daime, I began to sense, was more than a drug that induced an altered state; it was a plant that contained an essence considered sacred. Paulo refused to call it a drug. In fact, he claimed that daime cures addictions to alcohol and cocaine, that it never creates "bad trips" but comes from a healing tradition. It is taken mainly in ceremonies, usually once or twice a month. It is nonaddictive, purifies the system and is self-regulating, making one throw up if too much is taken. "If it's a drug, it's one that makes you encounter yourself, not run away," he said, "and if not for the daime, people would be drunk on *cachaca* [the white lightning of Brazil]."

(Later, in the United States, I learned that daime is made from *banisteriopsis caapi* and *psychotria viridis*. Dr. Andrew Weil, at the University of Arizona, confirmed most of Paulo Roberto's claims. He described daime as a powerful hallucinogen belonging to the family of indoles, which are widely found in nature. "Besides being present in the seeds of many legumes and mushrooms," Weil said, "there is a hormone produced in the pineal gland of our brain with the same structure, interesting because it suggests a commonality of all life." He said that daime is used by Peruvian tribes—who call it *ayahuasca*—for male coming-of-age rituals. He stressed, however, that daime has its desired effects in a ritual or religious context with positive expectations; taken by itself, the drug can set off adverse reactions.)

I asked Paulo Roberto if this was a cult we were visiting. With a laugh, he answered no, as if the question were frequently raised. Of the seven communities in the extractive reserve, only Mapia uses daime and, he said, "we are people you can count on. We get up with the sun and work until it goes down, and then we come here to pray. The sacred knowledge is freely taken. You have to seek it yourself. What is important is the program of action: economic, social and environmental.

"The forest is basic to our values," he continued. "If it is destroyed, we are destroyed. There was a time before man was a hunter when he was a kind of 'ecological man' who didn't interfere in the environment but lived in total harmony.

"What happened over a million years was an Oedipal problem between people and the Earth," Paulo Roberto said. "There was a splitting, as when a boy grows up from the parent. Since they couldn't have the father, who is God, they killed the father, then tried to possess the mother, who is Earth. So there was no longer a way of guiding people in how to live in nature. In the collective unconsciousness, there exists the memory of total harmony with nature, which the daime helps us discover."

I asked him if he was saying one needs the plant to really understand what he was saying.

"You can do it by yourself without the plant," Paulo Roberto replied, "but plants are better. God created them for our health, our bodies, but also for our souls. You need something to go inside yourself—it's like the deep sea; you cannot swim there by yourself."

Troy and I were exhausted. This was too much to think about. We agreed that we instinctively liked Paulo and the people and would let this adventure take its course. We went upstairs to a room we shared with a nest of bats. Troy quickly passed out, stretched diagonally in his hammock and covered by a sheet and mosquito netting. I looked out the window and was jolted by the sky. From horizon to horizon it seemed to be a thick glowing forest, a sky forest, of stars as densely packed as the trees below. When I closed my eyes, my mind expanded until I experienced the whole arc of sky and forest within me, and thought: "The universe in all its mystery may be centered here."

Mapia—January 11th

TODAY WE WENT looking for pink dolphins—that's right—in the Purus River and talked with visiting officials of the Brazilian environmental agency, one of whom wore a T-shirt saying "We are all entitled to make a mistake." The government representative led a meeting on how better to use the clearing, warning that beans take a lot of nutrients out of the soil. It was a relief to hear such familiar and practical discussion. But I can't get the talks with Paulo out of my mind.

Tomorrow will be the ceremony.

Mapia—January 12th

THE CEREMONY BEGAN like a country Sunday service anywhere. About 200 people came flocking in, the men mostly dressed in white jackets and pants, the women in white dresses with a green overlay, like vines, on their shoulders. It was the ninety-first or ninety-second birthday of Padrino Sebastiano, whose remains lay in a small white crypt by the church. I decided I would follow Paulo Roberto to the service, which he joined after some morning errands, while Troy continued sleeping (the bats had kept him up). I could hear a rhythmic chanting from the church several hundred yards away.

"How long will this be going on?" Troy had asked the night before. "All day, eight or nine hours," we had been told, which deepened the mystery.

In the center of the church was an altar of wood, a trinity-shaped star within a Star of David. The sun, the moon and the stars were represented as the trinity of nature, interwoven with Christian symbols and linked by the root of daime. I noted one thing truly weird, or funny: one of those battery-powered dancing flowers you pick up at airports. It was making the same movement as the people around me. The men (the sun) and women (the moon) were rotating round the altar in a shuffling two-step, singing hymns that contained the religious teachings. A high energy was maintained by stringed instruments and the shaking of cans filled with pebbles.

> I will call the star of water
> To come and illuminate me
> Give me strength, give me love
> Allow me to enter
> The depths of the sea
> The depths of the sea
> The force is the daime
> The daime is the Light
> He is the Messenger
> On the way of Jesus
> For ever I must love
> The sun, the moon, the stars,
> The forest, the wind, the sea. . . .

On two tables were urns of daime, which Paulo motioned me to drink if I chose. The matter was not simple. Since we live in an age of drug confessions, here's mine: I had never done any hard drugs, even though chemicals seemed to be exploding all around me in the sixties. I rejected the notion that inner change could happen by reliance on external substances, and I was afraid of losing control. I covered this anxiety with the belief that one should reach inspirational breakthroughs in a "natural" state (though I didn't mind the chemical of alcohol).

So why start in the nineties? If someone asked me to take daime in the Broadway Deli, I wouldn't be interested. But this seemed to be an opportunity that would come only once, to share an ancient ritual, in a tribal setting, in the wildest environment in the world. I was less afraid of ego loss at this point in my life than of missing this experience. To say no would have been like being among the Oglala Sioux 100 years ago and refusing a peace pipe on the grounds that I didn't use tobacco. Troy could make his own choice.

I reached for the daime, telling Paulo with a laugh that I wanted to be able to keep taking notes. "One glass then," Paulo said. "That will open

the doors and be enough. Two, three glasses"—he waved his hand—
"and you will go somewhere else."

The drink was bitter. After downing it quickly, I sat with Paulo
Roberto off to the side of the service. "Breathe in and stay with the
vibration of the music," he said. I remembered a question: Why would
the service take eight hours or more? "Because," he said deeply, "self-
purification is a long process, you have to reach your limits—physical,
emotional, mental, everything—before you can transcend them."

What was Paulo Roberto's dark side, I wondered. Why would an
educated shrink from Rio believe in the divine power of a plant—unless
it gave him power over people? I didn't have an answer but somehow
chose to trust him in this situation.

After a few minutes, Troy arrived, freshly washed. He looked at me
suspiciously, with a slight grin. "Did you take any? How much?" he
asked. When I told him, he asked whether he should. Paulo said a little
would be all right, and so Troy took two or three swallows. Father-son
bonding for real, I joked. He sat down next to me, taking pictures of the
dance.

The doors opened slowly. After about an hour, I felt myself becoming
more serene, floating calmly.

I looked at my son. He seemed centered, watchful, quiet. Smiling, I
asked him how he was feeling. "Nothing at first," he said. "But now the
dancers are starting to spin." A while later he turned with widening
eyes. "My legs just went *vooom* from under me. I can't tell if you're
hearing me or if my voice is only in my head." He muttered something
and laughed. I told him to just close his eyes and tell me what he saw. "I
see the forest everywhere," he said. After these two weeks it was not
unknown, fearful or claustrophobic anymore.

I looked at Troy and realized that I had been fearing a disconnection
with his leaving home. Was it the same disconnection that develops
between nature, the source of our life, and the human species it gives
birth to? Can that cycle be restored?

I kept taking notes while one after the other the symbols of the
universe flashed through my mind. The music lofted me. Then I surren-

dered myself to the power of the nature concentrated there, and the forest around us became a holy cathedral, full of song.

Los Angeles—January 15th

> I found myself back in the sepulchral city resenting the sight of people hurrying through the streets to filch a little money from each other. . . .　　　　　　　　　Joseph Conrad, *Heart of Darkness*

I'M BACK IN L.A., waiting for the Persian Gulf war to begin on television.

Here's a good question: Are we ultimately more dependent on rain forest resources than on Middle Eastern oil? Does anyone care?

Dr. Noel Brown at the UN cares. He called to discuss the possible Goodyear contract for the Mapia community. By 1992, 700 tappers there might produce and ship 1,000 tons of rubber here. More investments will follow, hopefully, from big medical and pharmaceutical firms. In the last two years [since 1989], twenty-one firms have begun importing rain forest products, including Ben & Jerry's (Rainforest Crunch), Patagonia (buttons) and the Body Shop (ingredients for cosmetics). Negotiations are going on with ninety-two others, including some *Fortune* 500 companies. Between Sting fans and Greenpeace members, there is a growing green market.

I've called Tom Lovejoy and spoken to University of California officials about new research ventures in the Amazon, and the response has been good.

Troy feels changed by his experience. It felt like we were two brothers living together, he said, not like a dad and kid. As for the daime, he thinks it would make one feel "totally paranoid around here" and strongly believes it should remain in the forest. I agree.

But could the spirituality, the inner journey and the environmental wisdom of the forest people be transplanted to this society of asphalt, lead, asbestos, oil and chemicals? Is harmony with nature an outdated idea in a world of industry and war?

The question, and the Amazon experience, create in me a vulnerability, a palpable pain, in response to nature's agony in the modern world. I can choose between the pain becoming unbearable or numbing myself enough to work on reform. Small steps are all one can take every day, I know, and an optimist would note that global concern for the rain forest has increased significantly this decade. But fifty acres a minute is no longer an abstraction to me. I see nightmare visions of it every day. I would not wish these visions on anyone.

The age of species extinction will not be ended soon and not by mere tinkering with industrial consumer society. It also means a new vision— Lovejoy's "real revolution," Incan wisdom, the rediscovery of nature in our innermost selves, environmental spirituality, call it what you will— that finally sees the evolving earth as our only living parent and deepens the vital connection as we did, father and son, in Amazonia.

The following books can provide further information on the rain forest:

The Burning Season: The Murder of Chico Mendes and the Fight for the Amazon Rain Forest, by Andrew Revkin (Houghton-Mifflin, 1990)
Fate of the Forest: Developers, Destroyers and Defenders of the Amazon, by Susanna Hecht and Alexander Cockburn (Routledge, Chapman and Hall, 1989)
In the Rainforest: Report from a Strange, Beautiful, Imperiled World, by Catherine Caufield (University of Chicago Press, 1984)
The Primary Source: Tropical Forests and Our Future, by Norman Myers (Norton, 1985)
Tropical Rainforests, by Arnold Newman (Facts on File, 1990)

Milken, Junk Bonds and Raping Redwoods

Bill McKibben

*I*t's always been pretty easy to understand how people acquire great fortunes, marvelous homes, beautiful paintings. John D. Rockefeller owned scads of oil wells; hence he was rich. Henry Ford sold a lot of motor cars, and Hearst a lot of newspapers. But the new wealth that flooded into Manhattan in the 1980s was more mysterious. The product of junk bonds, mergers and acquisitions and leveraged buyouts, it seemed divorced from oil or coal or grain or cement. Instead, it was described as the product of "restructuring" or "management reshuffling" or other occult processes. The twenty-four-year-old MBAs were certainly working long hours, and the dollars they were making were real enough—they could buy fancy dinners with them, or join health clubs. But it seemed like play money, collected by passing Go, unconnected with real life.

If you stopped to think about it, however, it was obvious that the money that paid for all those co-ops and car phones and cocaine had to come from somewhere; you could restructure Nabisco nine times over, but someone someplace had to be buying cookies. A good story to explain the link between the unreal and the real—a parable for the Reagan decade—has been taking place over the last few years in Humboldt County, California, about 250 miles north of San Francisco. It is a story about Michael Milken, junk bonds, Ivan Boesky, Drexel Burnham Lambert, enormous conglomerates—and between 10,000 and 12,000 acres of virgin redwood forest.

The 2,000-year-old redwoods are part of a much larger tract owned by the Pacific Lumber Company. For generations, PL, as the company is known, had been logging trees in the area—but cutting them fairly slowly, at a rate that would allow them to grow back, and usually cutting them selectively instead of simply clearing huge areas. Though the company owned the world's largest privately held stock of redwoods, it had very little trouble with environmentalists. In fact, over the course of seventy years, Pacific Lumber gave the state of California 20,000 acres of redwoods, helping the Save-the-Redwoods League at one point in the 1920s to acquire a huge parcel.

And Pacific Lumber seemed too good to be true in other ways as well. Wages were high; workers could rent cheap and tidy homes in the company-owned town of Scotia; every PL child got an $8,000 college scholarship. "This company was the elite of all companies," says Bob Younger, who works as a lumber handler at Mill A in Scotia and has lived in the area all his life. "They took care of everything, they looked after you, they treated you like family." Cast Jimmy Stewart for company president; it was a pretty wonderful life.

In early 1986, however, a Los Angeles-based conglomerate called Maxxam took over Pacific Lumber. It was a deal that summed up the decade. For one thing, Maxxam and its chief, Charles Hurwitz, had no particular expertise in the timber industry. Hurwitz's first big move had come in 1973, when he took over McCulloch Oil. Selling off many assets, he soon acquired the Simplicity Pattern Company; there, according to the *Houston Chronicle*, he sold off the company's famous dress-pattern line.

The Pacific Lumber deal, though, was a particularly avant-garde move. Calling on the cash-creating power of the Wall Street investment bank Drexel Burnham Lambert, Hurwitz borrowed $754 million to buy the public company—borrowed the money with junk bonds, the high-interest notes that have fueled the takeover binge of the last ten years. The exact details of the transaction are murky, but they have led to all sorts of revelatory allegations. A shareholders' suit, for instance, alleges that Hurwitz tempted the old board of directors with nifty golden parachutes. Complaints filed by the Securities and Exchange Commission [SEC] against Milken and Drexel Burnham allude to the Pacific Lumber deal, though as Maxxam spokesman Donald Winks is quick to point out, the company has not been formally accused of anything. There have even been charges by some shareholders—denied by Hurwitz—that he illegally parked Pacific Lumber stock, in an effort to conceal his moves toward acquisition, with Boyd Jeffries, who has admitted parking stock for Ivan Boesky.

The various charges probably don't bother Hurwitz too much—he has spent much of his professional life in the diverse legal wrangles that high finance so often entails. (According to *Fortune*, his first brush with the SEC came in 1971, not long after his graduation from college, when he, along with a number of his colleagues, was charged with plotting to inflate the price of a stock. He signed a consent decree without admitting guilt.) Amid the current welter of briefs and depositions, a couple of numbers stand out. One is twelve and three-quarters percent, the annual interest due on some of the junk bonds that financed the takeover. Another is $83 million, the total interest that comes due on the junk notes this year. And these numbers matter. Pacific Lumber was always profitable, but it never made the kind of money that would allow it to pay off a debt of that magnitude and still turn a profit.

Hurwitz, in other words, did not bring to Pacific Lumber some vast fund of knowledge about the woods, nor any great new ideas for cutting costs. He brought to Pacific Lumber a massive debt, which the company can pay off in only one way: cutting more trees. Since Hurwitz took over, the rate at which Pacific Lumber cuts its trees has doubled [as of 1989]. Because Michael Milken invented this wonderful new tool called the junk bond, twice as many trees are crashing to the ground in Humboldt County.

Pacific Lumber contends that debt service is not the only reason for the accelerated cutting: A timber survey that Hurwitz commissioned soon after taking over the company showed that it had more trees than previously thought, and so, says Winks, the loggers are engaged in "taking off inventory." And indeed, there's reason to believe that even if Hurwitz had been given the company for free, he might have increased the rate of cutting. In a meeting with Pacific Lumber employees, he told workers his version of the golden rule: "Those who have the gold rule."

Company president John Campbell calls Hurwitz's remark "unfortu-nate. He was trying to make a joke. It was an effort at making some humor." Public-relations man Winks says, "It was an ill-advised attempt to be humorous." But perhaps enlightening all the same: Compared with "Do unto others as you would have them do unto you," it's quite a contemporary aphorism. Hurwitz, says Winks, is "puritanical," uninter-ested in the perks of wealth, adding, "He doesn't have a jet; he flies commercial," which seems to be the current standard for beatification. He's "an exercise addict," and though he inherited a sizable sum from his dad, a shopping-center pioneer, "he's just a guy who's driven by the need to succeed in business."

And succeeding in business means making more money. "What you had here was a company with an enormous amount of assets being undermanaged," says company president Campbell. "It was sleepy," says

Winks. "It was like many American industries that operated that way and hence became targets." The slowpoke old company would eventually have chopped down the virgin groves (though environmentalists say there was at least a slightly better chance the pre-Hurwitz PL would have sold the land to the state or the feds). As a result of the invigorated Hurwitz management, everything is speeding up. Several hundred new jobs have been created, new mills built, extra shifts added.

Surprisingly, not all company employees are overjoyed with the developments; a number have banded together to try and buy out Pacific Lumber with an Employee Stock Ownership Plan, or ESOP. "They're cutting way too fast," says mill hand Younger. "Before, when you got a job at Pacific Lumber, you got a job for life. The timber was always going to be there. But now, if you ask me, it's all going to be gone in fifteen years. I'm twenty-eight—I'm looking ahead, and I see that I'm going to have to do something else."

Though Younger says as many as 200 of the company's 1,400 or so workers have backed the ESOP, "most of the people still believe PL will take care of you, because that's the way it's always been. They don't understand times have changed." At the moment, he says, he's hoping that Hurwitz, who (the ultimate eighties guy) was involved in a busted Texas savings and loan, will decide to sell. "We'd just like to see it get back to sustained yield, to where we can have jobs in the future," he says.

Probably no one would have paid much attention to the labor woes of one lumber company, though, if it hadn't happened to have owned the last privately held groves of virgin redwoods in America. "If this was Douglas fir or southern pine, no one would give a damn," says Winks. Instead these trees are coastal redwoods, many of them born before Christ. Listen to Greg King, a local environmental activist: "There's one grove in there that epitomizes it all for me. We call it Headwaters Forest; it's on the Little South Fork of the Elk River, which is probably the finest redwood stream in the world. It's totally intact for about two miles; it's like a little microriver. There's a tree I just found in there recently that's between fifteen and twenty feet in diameter. One gets the feeling of being dropped into a world totally apart from this planet."

King and other local activists have gotten to know the redwoods pretty well. On several occasions, most recently in early June, members of a self-styled radical environmental outfit called Earth First! staged sit-ins in the upper reaches of endangered redwoods; to fell the trees, loggers would have had to fell the protesters. Activists from Earth First! and other groups have won lots of publicity across the West in recent years by destroying bulldozers, pulling survey stakes to slow down road building and other "ecotage" efforts. (They've also won the attention of the FBI, which arrested Earth First! founder Dave Foreman this spring on charges that he conspired to tear down power lines.) In Humboldt County, though, their protests have been nondestructive—classic civil disobedience. And their leaders are hardly wild-eyed mountain men.

Darryl Cherney grew up in the forests of Manhattan and first saw the redwoods the way almost everyone else first sees them—from the back of his parents' car. "I'd always had this romantic notion of moving back there to live," he says, "but I also had this New York notion that to live someplace that beautiful must cost several thousand dollars a month." By the time he finally returned to northern California in the mid-eighties, the Pacific Lumber groves were being threatened with destruction, and so he has spent at least as much time hiking through clear-cuts as idyllic forests. "In the places they cut old growth, they've announced that they'll leave a few trees," Cherney says. "So it looks kind of like a dirt parking lot with an occasional tree sticking out in the middle. No ferns, nothing—just a tree." Because of their isolation, Cherney says, many of the remaining trees have been toppled by the wind.

The Earth First! activists have joined a number of other environmental groups seeking to block the old-growth cutting not only in California but throughout the Pacific Northwest. In a series of lawsuits, environmentalists have argued that a number of wildlife species, including the spotted owl, the red tree vole and a shorebird called the marbled murrelet, depend on the snagged tops and shady bottoms of old-growth forest. Maxxam spokesman Winks says that "since owls have wings, if the trees are gone, they can fly a few miles and find some more." But the courts, at least on a few occasions, have sided with the birds, temporarily halting some logging.

The environmentalists also dismiss the company argument that plenty of redwoods are already protected in state and national parks. "They say that there's 80,000 acres of old growth protected," says Cherney. "But a lot of it is Winnebago park, hundred-foot-wide strips." In some places holes have been cut in the trunks of the redwoods so paying customers can steer through: drive-in nature. Overall, only about three percent of America's virgin forest remains unlogged. "That's just not enough," says Greg King.

And so some environmentalists are pushing for state ballot initiatives in 1990, calling on California taxpayers to save all the islands of old growth. If the propositions are passed, foresters won't be allowed to cut in a grove till a certain percentage of trees have reached 150 years of age.

At root, the dispute between Maxxam and the environmentalists is about different ways of looking at wilderness, different ways of thinking about man and earth. One view, just beginning to emerge, is expressed by King when he talks about hiking through an endangered grove. "It's hard to put into words what I feel except to say that it is life—it is reality," he says. "These trees are good friends of mine, as dear to me as any person."

The other, more conventional, view has been with us since some Stone Age fellow made the first ax. I asked spokesman Winks if Maxxam chief Hurwitz had ever seen these ancient groves, and if so, what he thought about them. "He's visited the facilities he now owns, but he hasn't made himself an expert," Winks says. "His interest is in seeing that the company set business goals and that they meet them. I think that he believes that there are lots of redwood trees, and that these are the ones he's paid for and now owns. . . . It's hard to argue with the emotional argument that an old tree is special. But there are more pragmatic arguments. An old tree takes up space that could be used by more and younger trees; a big tree produces good revenues. And it's like any other piece of property. What about his constitutional rights?"

Perhaps Hurwitz and Milken and the rest are correct—perhaps what really counts is that people keep building redwood decks and hot tubs (two of the prime uses of the lumber) and that jobs are created as quickly as possible and that men who are already rich grow richer. But I've seen both the canyons of Wall Street and the fog-hung cathedral forests of the Pacific Coast. And so it's hard for me not to wonder if King, the tree hugger, hasn't reached some profounder understanding: "We've degraded so much of the planet," he says, "that reality seems unreal."

PART IV

Anatomy of a Disaster

With the environment, it is often said that people will not believe that catastrophe is coming, but will only try to protect themselves after an accident occurs. If this is the case, then certain accidents stand out above the rest in the mark they have made on human consciousness.

Chernobyl, Three Mile Island, Love Canal, Bhopal and the *Exxon Valdez* are among them. All could have been avoided, and yet all occurred. All pit the magnitude of our use of resources, production of wastes and demand for consumption against the potential fragility of the environment. And all question our priorities, uncovering a willingness to risk irreparable and fundamental environmental losses.

These cases offer some of the most visual and disturbing evidence that mistakes with the environment are costly, perhaps more costly than we can afford. Tom Horton's exploration into the making and the aftermath of the *Exxon Valdez* disaster is a chilling reminder of the causes and effects of environmental catastrophe, effects that play out long after the networks' week-long, even day-long, coverage ends.

Perhaps, in this case, even more chilling than the sight of wildlife drowning in oil and human livelihoods destroyed was the indecorous task of assigning blame and collecting restitution for the disaster. The case of the *Exxon Valdez* was particularly complex in the messages it sent to the public and to corporations. Since this article was writ-

ten, Joseph Hazelwood, captain of the *Exxon Valdez*, was
acquitted of all criminal charges, except negligence, which
is still under appeal. A first settlement in the suit against
Exxon was eventually replaced by a second settlement that
made Exxon pay more. Many spectators, however, claimed
that Exxon received tax credits that more than covered the
cost of these penalties.

The Oil Pollution Act of 1990 was part of the legislative
aftermath of the disaster. It raised the operational standards
of oil companies, requiring equipment improvements like
double-hull tankers and raising the liability limits to
which a company can be subject. But the Oil Pollution Act
has been highly criticized by environmentalists, many of
whom think that it provides no real penalty to Exxon or
any substantial incentive to avoid accidents.

The only real cost that came to Exxon, they say, is that
the company now has to pay much higher insurance pre-
miums. (The insurers raised their rates to cover natural
resource damages.) That, not the lawsuits or changes in
public opinion, the argument goes, is the only thing that
may have gotten to Exxon.

The results of the *Exxon Valdez* spill are difficult to measure. Estimates of the number of birds killed span a wide
range, and few people believe any of them. Pressure to
open the Arctic National Wildlife Refuge continues in
Congress and in the Bush administration, despite the blow
that the *Exxon Valdez* incident presented.

Even with such unclear results, it is important not to
forget accidents like the *Valdez*. Pressures will build, and
consciousness will change. As long as environmental catas-
trophes do not slip out of public debates and forums, they
can gradually erode away the structures that make them
possible.

Paradise Lost

Tom Horton

*I*t's 4:30 A.M. when my ride to Bligh Reef, the supertanker *Arco Prudhoe Bay*, slips serenely from the port of Valdez, Alaska, her belly freshly gorged with 50,000 tons of hot crude oil from the North Slope. There has been a slight addition to normal sailing orders, says Captain Justin C. Raymond, a former lobsterman from Nahant, Massachusetts: "Keep her off the rocks, Dad," his fifteen-year-old son has urged him. Raymond's son has been taking flak from schoolmates since another tanker captain, Joseph Hazelwood, crunched the *Exxon Valdez* head onto the reef during a similar transit through Prince William Sound in March [1989], nearly a month ago now. "I've tried every way I can since then to imagine how they did it," says Raymond. "It was just insane."

The captain, wearing jeans, Nikes and a parka, sips coffee at the big glass windows of the bridge. Dawn is slopping over the snowy peaks of the Chugach Range, which walls our passage on both sides. From the airy, spacious bridge, done in light browns and cool creams, the scenery, not the tanker, seems to be moving. Only the merest shudder betrays the workings several decks below us of 20,000-horsepower engines, chuffing along at a lazy eighty revolutions per minute, turning a propeller roughly the height of a two-story house. Commanding all this is the most disappointing steering wheel, a black metal disk about eighteen inches in diameter (hydraulics have replaced the big, polished wood and brass affairs, which would, for a supertanker, have to be about thirty feet high to afford sufficient mechanical leverage on the mammoth rudder). All in all, the ambience is more insurance-company office than salty fo'c'sle; Metropolitan Life, not *Tugboat Annie*. There is no sense, here in the calm of Prince William Sound, that we are riding a potent reservoir of toxic cargo that, for all its impressive propulsion and guidance systems, is only marginally maneuverable.

Anticipation is all, in driving one of these babies. For example, our bow, sticking out there farther from us than Bo Jackson's longest homer, is pointed dead on for a granite slab of Chugach mountain. Not to worry; Jim Wright, one of the local pilots every tanker must carry out of Valdez, called out a course change to the able-bodied seaman at the helm nearly a minute ago. It will be another half a minute or so before we can see the bow of the *Prudhoe Bay*, 662 feet distant, clearly responding, and yet another minute before the AB calls out that she has steadied on the new heading.

A plaque on the port side of the bridge details the results of the *Prudhoe Bay*'s latest deceleration trials, which demonstrate what happens when she loses power while underway. Running at full sea speed, about fifteen knots, her engines were thrown into idle. Forty-five minutes later, having traveled 7.4 nautical miles, she was still plowing along smartly at more than four knots. Full speed astern to slow her, says Raymond, is a good command in the movies, but of scant utility in real life. It only causes the huge prop to "walk" a supertanker's stern around, sending her careering uncontrollably. "Strictly a desperation move," he says. "The only way to bleed off speed is a series of hard turns."

It's 7:00 A.M., and we have cleared the Valdez Narrows, a pinched passage through the mountains where tankers are restricted to six knots. Captain Raymond is kidding Pilot Wright that it has just taken the latter fifty-two seconds to visit the head. "Some expert" from the California Maritime Academy, says Raymond, has been timing people recently, just a small example of the almost frenzied theorizing that is going on to explain why Captain Hazelwood disappeared from the bridge of the *Valdez* not long before she foundered on the rocks. "You should have been able to do it in forty-four seconds," says Raymond. "Well, I wash my hands after," says the pilot. "Better with the old ships . . . you just pissed on the deck," says the captain.

How did the *Exxon Valdez* wander from a shipping lane so wide that a United States Coast Guard admiral would later say, with only mild hyperbole, that "your children could drive a tanker through it"? As we near the rocky underwater promontory that was the *Exxon Valdez*'s downfall, the mystery seems only to deepen. These ships have so many navigation systems: Besides twin radars, there is Loran C, a sophisticated triangulation system that uses shore-based signal towers to direct a boat to within several meters of any spot it has been, even in open water. A separate system, SATNAV, which relies on satellite signals, does much the same as Loran C. Then there is the Collision Avoidance System, CAS II, basically a minicomputer with a screen that gives clear and detailed visual information of the ship's speed, course in tenths of degrees, closest point of approach to other objects and time of arrival at any point on the screen.

There is more technology to see, but an old navigation phrase, the lowest of low tech, has been running through my mind ever since we left Valdez: RED RIGHT RETURNING. My father drilled it into my head when I was twelve, running a fifteen-foot skiff with ten-horsepower outboard around the shallows of my native Chesapeake Bay. RED RIGHT RETURNING. Always keep the red channel markers and red buoys on your right when returning to a port or harbor, and conversely, keep them on your left when you are leaving. RED RIGHT RETURNING. It's information good for getting in and out of any established navigational channel in America, and much of the world. The way in and out of Valdez is no exception. Virtually all but the most casual pleasure boaters know the RRR dictum. It certainly must have been known to the crew on the bridge of the *Exxon Valdez* that fateful morning, a few minutes after midnight, March 24, 1989, when they apparently ignored it, to the everlasting regret of us all.

It was a fine night for sailing, calm and clear, when the *Exxon Valdez*, the 987-foot flagship of the giant oil company's navy, sailed at nine.

Fishermen around Valdez would recall afterward that the northern lights that week had been unusually active, flaring a rarely seen reddish orange, and maybe it was an omen. Thursday night, even as the ship left port, the people of Valdez were meeting to form a committee to deal with the impact of oil transport on the area, and Riki Ott, representing one of the fishing organizations on Prince William Sound, testified that a major spill was "a question of when, not if," but she was known for being outspoken.

Joe Hazelwood may or may not have been drunk at the wheel. He has been fired by Exxon and is charged with three misdemeanor and three felony violations, carrying a maximum sentence of five years and $50,000. He is not talking to investigators. We know secondhand his step-by-step movements ashore during most of the preceding day—shopping for Easter flowers to be sent to his family in New York State, ordering pizza, drinking what appeared to be vodka, playing darts at the Pipeline Club, a popular tanker men's spot. After returning to the ship, he had two nonalcoholic Moussy beers, he told a Coast Guard officer who arrived on board soon after the grounding. Not until nine hours after the accident was the captain tested for blood alcohol. He was found to be legally drunk—but by then, who in his place wouldn't have been?

Hazelwood's actions that night, according to witnesses at government hearings and interviews with crew members, were strange at the least and violated accepted tanker practices. He left the bridge during the critical passage through the Valdez Narrows; left it again after ordering a course change to avoid small icebergs that set the ship on a line straight for Bligh Reef. He also turned the bridge over to Gregory Cousins, his third mate, who was not licensed by the Coast Guard to pilot the ship in these confined waters. For all that, it appears likely Hazelwood would have gotten away with it had Cousins followed the captain's orders—or just kept on the proper side of a clearly visible red light.

It was shortly after 11:39 P.M. that the *Exxon Valdez* began its death dance across the dark, smooth waters of Prince William Sound, based on testimony by the crew before the National Transportation Safety Board (NTSB). At that time a navigational fix taken by Cousins showed the vessel squarely in the middle of the established shipping lanes. Hazelwood ordered a course due south, 180 degrees, which would take the

Exxon Valdez across the shipping lanes and straight toward the reef. Before leaving the bridge, he and Cousins jointly agreed upon a point in the vessel's new course where the mate would hang a right, turning back to the west when he was three miles north of Bligh Reef, well in advance of any danger of grounding on it. Inexplicably, the captain placed the ship on autopilot, something mariners say is never done in such confined waters. It would be a bit like putting your automobile on cruise control while entering a parking garage. The mate or the able-bodied seaman at the wheel, Robert Kagan, soon returned the ship to manual, however. It does not appear, as was once thought, to have been a factor in the crash that followed.

RED RIGHT RETURNING. Maureen Jones, the lookout on the wing platform extending outside the bridge, was perhaps the first to know something was badly amiss. At about 11:54 P.M., according to the NTSB testimony, she noticed a red flashing light well to starboard. Red lights when leaving harbors are supposed to be on your port, or left, not your starboard. She strode quickly inside and reported it to Cousins. He made a calm, routine acknowledgment and returned to his chart work.

Jones went back to her position, checked the red flashing light, now moving even farther off the ship's starboard. She identified it positively as the Bligh Reef light by its flashing sequence of once every four seconds. Again she entered the bridge and reported to Cousins that he had a red on his right. Again he made a routine acknowledgment as the giant ship plowed due south. The lookout again took her position on the bridge wing, noting that the bow seemed to be coming slowly around to the right. Cousins would later testify to the NTSB that he ordered the ship turned to the west at about 11:56 P.M., in what should have been time to avoid the reef. Automatic recording instruments aboard the *Exxon Valdez* showed, however, that the turn was not begun until 12:01 A.M., three minutes before grounding. Robert LeResche, a state of Alaska investigator participating in the NTSB hearings, likened this five-minute gap to the mysterious eighteen-minute gap in Richard Nixon's White House tape recordings during Watergate.

LeResche asked Cousins about the discrepancy. "I really don't have an adequate answer," said the mate.

A few seconds after Maureen Jones noted the bow responding, the *Exxon Valdez* bulldozed the reef, which until then was best known to Alaskans for its good halibut fishing. The vessel's mass of hundreds of millions of pounds was so great that the impact was surprisingly gentle. Jones saw a glowing aura around the bow for about fifteen seconds. Some on board were not aware they were no longer moving. Second Mate Lloyd LeCain, asleep after a long shift, was awakened and saw an assistant engineer opening sounding tubes in the deck to check if water had entered dead air spaces built into the ship's hull. LeCain saw a geyser of crude oil shoot an estimated seventy feet from the tube into the air. Chief Mate James Kunkel knocked on LeCain's door and told him, in the inimitable wording of the NTSB report, "Vessel aground, we're [expletive]." At 12:27 A.M. the Coast Guard radio in Valdez crackled with Joe Hazelwood's voice, reporting the grounding and, in the understatement of the day, adding, "Evidently, we're leaking some oil."

And off to starboard, the right side but the oh-so-wrong side, the red flashing light of Bligh Reef continued to blink its impotent warning every four seconds, and 11 million gallons of brownish black crude oil, North America's biggest spill, began spreading across North America's greatest wilderness. In the dim glow of the nighttime bridge, Alaska State and Coast Guard investigators would find Hazelwood, when they boarded the vessel a couple of hours later, "looking pensive." Almost gently, one of them asked the captain if he didn't think he should snuff his cigarette—since the fumes might be explosive. "Oh, yeah, you're right," Hazelwood reportedly said.

Back on the *Prudhoe Bay* it's full daylight, Bligh Reef is slipping comfortably astern, and it's time for Jim Wright and me to get off or else travel another five days with the tanker to Long Beach, California. The ship is moving at fifteen knots, in moderate ocean swells, as we clamber down a rope ladder hung from the Arco supertanker's side onto a tiny pilot craft that keeps pace alongside and far below the ship's deck. Twenty minutes later the long morning spent in the modern commerce of oil seems a distant memory. I am surrounded by eagles and on the lookout for bears. The pilot boat has put into a rock-girt cove about fifteen miles out of Valdez, where an old yacht is moored as living quarters for the men who embark and disembark the pilots. Fresh drinking water arrives via a garden hose run ashore and stuck into a waterfall that ripples down the hemlock slopes. Bear and deer routinely travel the shoreline within sixty feet of where we are moored; also mink and coyote. The men here fish in their spare time, catching red snapper, grayling, cod, sea bass, halibut and Dolly Varden.

Do I see many eagles on the Chesapeake Bay? asks Matt Ortega, one of the pilot-boat operators. A few a year if I'm lucky, I tell him. They're coming back from the DDT that destroyed their eggs when it got into the food chain. He walks on deck and whistles sharply. On the second blast, six adult bald eagles come winging across the treetops. Ortega throws a piece of bologna into the crystalline water of the cove. No takers. "Ah, can't fool them," he says, ducking inside and reemerging with a pound of bacon. "They like that bacon." For the next fifteen minutes the great predators circle in formation as he throws strip after strip into the water no more than ten feet from the boat: circling, swooping, gliding, gliding, gliding, powerful legs swinging slowly forward, talons extending. You are almost hypnotized, a crouched rabbit awaiting death, then—choonk!—the bacon is ripped from the water with savage precision, scarcely rippling the surface.

You could do that trick in the lower 48, I tell Ortega, but instead of bald eagles, mostly you would draw hordes of shrieking, squabbling herring gulls, so common we call them bay buzzards or ocean rats. Here is simply nature on a higher plane than exists anywhere else. You could spend a lifetime appreciating and enjoying that cove, and it is but one small indentation, a pinprick among thousands of miles of coves and inlets and embayments that are the shoreline of Prince William Sound.

The sound was first known to white men through the voyage here of Captain James Cook in 1778. A reef charted by his expedition was named after one of Cook's officers, William Bligh, who would later become identified mainly with more southerly seas and a mutiny on his ship the *Bounty*. Cook, seeking a northwest passage, was bound to be disappointed, but in 1899 the Harriman Expedition, an extraordinary voyage to Alaska that included America's leading scientists, naturalists and wildlife painters, brought the region's stunning beauty to public attention.

John Burroughs, the naturalist and author, wrote on entering the sound of "the vast shifting panorama of sea and islands and wooded shores and towering peaks spread before us on every hand . . . a feast of beauty and sublimity. . . . We were afloat in an enchanted circle." And his colleague, John Muir, who founded the Sierra Club in 1892 and was no stranger to alpine beauty, called it "one of the richest, most glorious mountain landscapes I ever beheld—peak over peak dipping deep in the sky, a thousand of them, icy and shining . . . and great breadths of sun-spangled, ice-dotted waters in front . . . grandeur and beauty in a thousand forms awaiting us at every turn in this bright and spacious wonderland."

For weeks afterward, whenever I read the daily Exxon press releases on the spread of the spill, on how many miles of beach had been tarred by crude, I would think of my visit to the eagles' little cove. Its total shoreline, had it been oiled, would have seemed unimpressive in the context of the sound's total, a tragedy expressed in hundreds of yards, no more; yet in reality so much more. Similarly, reports that no more than ten or fifteen percent of the sound's shoreline had been harmed by oil were accurate as far as statistics went, but it seemed like telling someone only a portion in the center of a rare and treasured painting had been ripped—cheer up, the rest of it was untouched.

The spill has painfully underscored a dichotomy that underlies much of the modern Alaskan experience. Prince William Sound until this March was exemplary of the Alaskan slogan "America's last frontier." If Americans were granted a new start, a chance to reinvent their continent's pristine landscape and its wildlife, its air and water, and to treat them right this time around, this would be the result. If it's not God's country, a resident told me, then God should homestead a piece of it while he still can. Clearly, Alaskans feel wilderness symbolizes their state. The beasts and fowls and fishes are on display in airports, motels and restaurants everywhere you travel—stuffed, mounted, painted, carved, silk-screened on T-shirts, lynxes, wolves, eagles, bears ten feet tall, leaping salmon, halibut the size of your front door, beavers, otters, moose, caribous, musk oxen, fifty species of waterfowls and wading birds—and here they are symbols, not of the past but of the present and, it still seems reasonable to think, of the future.

But it is not fur trapping or deer hunting or bird-watching or even its substantial fishing industry that runs Alaska. It is oil. Perhaps that could be said for much of the developed world, but here the focus is sharper, defined by a single pipeline, four feet in diameter, 800 miles long. From the pipe's mouth in Valdez, where the supertankers say fill 'er up, issue some 2 million barrels a day of crude. It comes out still steaming, only slightly cooler than it issued from the earth, an effect from the friction of being pumped from the frozen tundra of the North Slope across three mountain ranges, the mighty Yukon River and the great migration routes of caribous. The pipe took 70,000 people four years and $8 billion to build—America's largest private construction project (though only Japanese mills could deliver the quantity and quality of steel needed for the pipe's half-inch-thick walls). From the air the pipe's odd zigzag configuration, scientifically designed to permit thermal expansion and absorb even the shocks of earthquakes, appears as man's lone, bold essay on a natural landscape as fierce and remote as any on earth.

Two million barrels, 84 million gallons a day, flowing as sure as a river since 1977—the pipe carries nearly a quarter of all the oil produced in America, an eighth of the nation's total oil consumption. It is not stretching things much to say that one day of every week, the entire United States, from its automobiles to Du Pont's polyester production, is dependent on this one pipe.

The pipe, politicians tell people here, is worth the revenue of thirty gold mines; better than sixty molybdenum mines—worth such sums it almost seems that Exxon, one of the pipe's major owners, could as well use dollar bills to sop up the spill instead of the company's often ineffective booms and skimmer vessels. Profits conservatively estimated at $42.6 billion, after taxes, have come to the oil industry in Alaska as of 1987, almost all of it from producing and transporting oil from the North Slope through the pipe. The figures, which put the industry's after-tax return on investment at nearly forty-four percent, come from a study by Edward B. Deakin, a well-regarded professor of petroleum accounting at the University of North Texas. His study was commissioned by the Alaska Department of Revenue. The average profit on each barrel that has passed through the pipe since 1977, Deakin puts at better than $6. At recent flows through the line, that works out to around $12 million a day, or close to half a million dollars an hour, twenty-four hours a day, 365 days a year.

Neither have the taxpayers come off shabbily. During the same period, oil extraction from the state paid $25.8 billion in federal taxes and $29.3 billion in taxes and royalties to the state of Alaska. Oil generates a whopping eighty-five percent of all state revenues. A portion, invested mostly in bonds, has created the so-called Permanent Fund for Alaska, whose principal is projected to top $10 billion by 1991—rainy-day money of about $20,000 for every man, woman and child now living in the state. Alaskans, courtesy of oil revenues, enjoy generous college loans, a highly touted school system in which teachers start at $30,000 a year, a level of government services about sixty percent above the national average, subsidized public transportation . . . the list goes on and on.

The oil industry has been attentive to its public image. Boy Scouts get support from British Petroleum America; Exxon and Arco executives are active in well-publicized civic and charitable activities in Anchorage; Alyeska, the consortium of companies that operates the pipeline, has set records with its United Way contributions (the highest in the United States per employee in 1986 and 1987). And lest Alaskans not appreciate all of this, the industry runs television ads that come on with upbeat music while pictures of Alaskans in varying occupations flash across the screen. Each sequence of workers is followed by "Alaskan jobs"; then another sequence and "Alaskan jobs"; then the punch line—"depend on oil." "It's really obnoxious and really true," a Sierra Club lawyer in Juneau said of the ad.

There is more. Alaskans pay no state income tax, and once a year, courtesy of the Permanent Fund, or Big Oil, if you will, the state pays each of its half a million or so full-time residents money just for being here—a dividend that this year will total $862 per capita.

And yet . . . is this gratitude?

"We did not expect sympathy, but I think our people were unprepared for the fury and extent of the public reaction . . . the hate calls we got after the spill," said Joan McCoy, who works in public affairs in Anchorage for Arco, one of the pipeline's major owners.

But consider what a powerful and attractive illusion was shattered by the oil spill. This place has been having its cake and eating it too on a grand scale—Alaska the beautiful, Alaska the wild and pristine, coexisting, pretty companionably it seemed, all these years with Alaska the oil state, Alaska the flush. The postspill images seem less flattering: "If Alaska is not our young whore, what is she?" wrote Harry Crews in his *Playboy* essay "Going Down in Valdeez," written during the pipeline-construction boom more than a decade ago. "She is full of all that will pleasure us [but] if we scar her . . . who can blame us? Didn't we buy her for a trifling sum to start with?"

I thought Alaskans a bit hypocritical but mostly just naive in their libertarian disregard for lower-48 environmentalism, in their feeling that because they care so deeply about their outdoors, they are therefore good stewards of it. Indeed, they have, to an unparalleled degree, enjoyed their cake and good eating too, but that seems more a matter of a very big, unspoiled place with as yet only about half a million people nibbling at it. The unfocused, ugly sprawl around Anchorage, for example, exemplifies every unfortunate trend in land use that has trashed so much of the lower 48; but as yet the mountains are vast, the population is tiny, and wild moose still stroll into town.

So, yes, there is rage directed against the oil companies but also, perhaps, against themselves among Alaskans. "It's aggravating to me and others who should have realized all along the oil industry is here for profit—and nothing wrong with that," said Gregg Erickson, an economist in the office of Alaska's governor, "but we let ourselves be seduced, we forgot they weren't here primarily to support arts and Little League and United Way."

At first, I was surprised that Alaskans hadn't seized more ferociously on the man most directly responsible for all that happened here and made Joseph Hazelwood a whipping boy. But as the details have emerged of how the spill happened and subsequently spread out of anyone's control, it has become clear that the tragedy in Prince William Sound amounted to an indictment of the whole system. Joe Hazelwood gets no one's sympathy, but he was merely the trigger for a disaster that had been waiting to happen for years.

By 3:00 A.M. of the Good Friday spill, its terrible dimensions were becoming apparent: ". . . 138,000 . . . and that was barrels, correct?" a glum-sounding Coast Guard officer said as he talked with the bridge of the grounded ship (the estimate would later be upped to 240,000 barrels, about 11 million gallons). The first line of defense in containing and cleaning up the oil was the Alyeska Pipeline Service Company, whose name, Alyeska, is an Aleut word meaning "great land." It had been more than a decade since Alyeska, building the pipeline across the wilderness, had been called upon to perform heroic feats.

As Alyeska's crews labored in the predawn hours of March 24 [1989] to load pollution-fighting gear on tugs and barges, one wonders whether any of them looked for inspiration at the heroic Malcolm Alexander statue nearby commemorating the rough and ready spirit that built the pipe that transformed Alaska. A rugged, thirteen-foot-high bronze, mounted on a base of boulders, it features a woman teamster, a native workman and three other men clad in oilskins, boots and parkas, carrying the tools of the welder, surveyor and engineer. Their eyes fixed on a far horizon, they personify what was a favored company slogan of the construction era. "We didn't know it couldn't be done."

These days the pipeline and Alyeska's huge oil-loading terminal are mostly operated by a different sort—trained technicians who work in an orderly, predictable environment, supervising softly glowing computer screens, tending to the valves and flow meters and assorted plumbing of this larger-than-life filling station. They dress neatly, go home to families and immaculately kept houses and generally enjoy the good, settled middle-class life of Valdez. During a tour of the terminal, it struck me that one could work here for a lifetime and seldom, if ever, see, hear or smell the crude oil that flows in such immense quantities to the lower 48.

A few weeks after the spill, Tom Brennan, Alyeska's public-affairs officer, is explaining why he feels the company has taken an unfair beating for its failure to begin promptly cleaning up the oil slick before it headed for the pristine coasts of the sound. "I know people didn't see what they expected to see," he says. "I myself flew over at 8:30 A.M. [about eight hours after the grounding], and yeah, my first reaction too was . . . I expected to see booms surrounding the ship, but that just wasn't what was called for. No one envisioned anything like that happening. . . . It was obvious we weren't going to be able to contain it. . . . We knew that from the first." In other words, I thought later, *they knew it couldn't be done.*

Okay, that's unfair, comparing the pipeline construction to an epic catastrophe that happened in the middle of the night; in fact, weeks later it is pretty much the consensus of both government and industry experts who have worked other big oil spills that no existing technology could have quickly corralled and recovered substantial amounts of the oil. Finally, almost lost in the hoo-ha over the oil industry's incapacity to clean up the oil that leaked, is the pressing need that existed to unload the oil—four times what spilled—still aboard the *Exxon Valdez.* It was possible that the ship might break up and sink, and there is wide agreement here that lightering, or off-loading that oil onto other tankers, was at least as high a priority as cleaning up the spill. This task was accomplished without major delays by the Coast Guard and by Exxon, which quickly took over spill cleanup from Alyeska after the accident.

But for all that, you must grant a tankerful of skepticism about Alyeska's rationalizations to anyone who has read the company's contingency plan, the official document that details how the company would deal with any accidents in transporting oil. It was what the company had pointed to over the years—with a confidence that could border on the patronizing—whenever critics wondered what would happen if a supertanker ever cracked up in Prince William Sound. The "CP" is a formidable document, written with enough techno-babble about "boom deployment angles" and calculations of "oil encounter rates" to numb the layman, but page 1 is straightforward enough.

"The resources of Alyeska Pipeline Service Company are organized in a preplanned manner to ensure rapid and effective response to any oil spill emergency," it reads. No ifs, ands or buts. No "we never envisioned anything like this." In fact, a similar accident came near to happening when the supertanker *Prince William Sound* lost power and steering in high winds in 1979. The tanker was within half an hour of being blown onto the rocks when a wind shift brought it back into deeper water and it regained power. The only apparent response was to include better arrangements on board tankers for tugs to hook lines.

In 1982 the state ordered Alyeska to revise its cleanup plans to reflect a catastrophic spill of 200,000 barrels (the old plan envisioned a maximum spill of 74,000 barrels). Alyeska appealed the decision, and it was not until May 1986 that the company agreed to add the new scenario. The plan, after emphasizing that a 200,000-barrel spill was "highly unlikely," stated that the company would respond to spills in the area including Bligh Reef within two to five hours and that fifty percent of the oil would be recovered (some of it, the plan admitted, after it had hit shore).

In fact, nearly fifteen hours went by after the [1989] spill before Alyeska reached the scene with even token amounts of cleanup equipment. Several times during this period the company responded to state and Coast Guard inquiries as to when its crews would arrive by saying, "We're on the way." A truer picture of what was unfolding at the terminal that night was one of confusion and woeful unpreparedness, according to interviews with Alyeska employees, published news reports and testimony at congressional hearings about the spill. Equipment was buried under tons of boom deep in warehouses; some was hidden under several feet of snow. A barge that was supposed to be loaded with spill equipment wasn't. It had been damaged and, still seaworthy, was unloaded while awaiting repairs. Alyeska pointed out at one hearing that, technically, the barge wasn't required to be loaded. "But didn't common sense require it to be?" asked an exasperated congressman. For several hours the company had only one man on the scene to operate both a forklift bringing equipment to its barge and a crane lifting it onto the barge. The employee would have to jump back and forth between running the forklift and the crane.

There was also serious confusion between Alyeska and the Coast Guard, the agency with overall authority to direct the cleanup. At one point, this resulted in the company's beginning to off-load spill-response equipment from its vessels so it could load lightering equipment, something the Coast Guard says it never asked the company to do. Then there was the fear on Alyeska's part that booming so much crude oil in the vicinity of the crippled tanker carried the risk of massive explosion from the concentration of fumes coming off the slick. It is still not clear whether such concerns were groundless or not, but nothing was ever said about such a possibility during the months of detailed discussions and review that went into developing the contingency plan. An Alyeska employee who was at the center of the response effort that day, and who did not want to be identified, said of the contingency plan, "Whoever wrote it . . . well, all I can say is that they never had to respond to a 200,000-barrel oil spill."

For the next two days the weather was sunny and dead calm—unusual in March, when gale-force winds and snow are not uncommon. During that time it became painfully apparent how inadequate was the cleanup equipment Alyeska had on hand in Valdez. Commercial fishermen, who stood ready to throw dozens of boats into any oil-recovery effort, met total frustration. "They never returned one of our calls," said Jack Lamb of the Cordova District Fishermen United (CDFU), which represents Cordova, the major fishing port on Prince William Sound and the seventh biggest in the United States in value of catch. The oil, in the words of a Coast Guard officer, "just lay there," several square miles of it, seeming to mock the testimony an Alyeska executive, Ivan L. Henman, would later give to Congress: The contingency plan was carried out in a "timely manner . . . equipment operated as expected . . . crews performed admirably."

Beyond the traditional cleanup techniques of encircling the oil with booms and collecting it with skimmer craft, there may have been another way to prevent the *Exxon Valdez's* cargo from fouling hundreds of miles of coastline. Spraying dispersants from airplanes could have helped dissolve the oil slick before it hit land. Because dispersants are virtually the only tools for dealing quickly with huge spills, because Prince William Sound had one of the nation's best plans for using them and because the plan failed so miserably, it is worth examining the bitter political battle that has arisen. It will likely be fought again wherever a massive spill occurs in an environmentally sensitive area.

The first time dispersants were ever used on a major oil spill was after the sinking of the tanker *Torrey Canyon* off the Isles of Scilly in 1967. "They didn't know what they were doing then and used stuff so toxic it never should have been put in the marine environment," said a retired Exxon chemist who did not wish to be named. But dispersants have evolved considerably since then, he argued. That view was substantially supported by the National Academy of Sciences in a 1989 report on the current generation of dispersants. The academy said they were no more toxic than the oil itself.

Dispersants are not without their disadvantages, but they suffer from a reputation that inspires undue fear and misunderstanding, according to experienced oil-spill experts of the National Oceanic and Atmospheric Administration (NOAA). Dispersants shatter oil's natural cohesiveness, breaking it up into tiny droplets that no longer float as a slick on the surface but disperse throughout the water. This allows rapid degradation of the oil, through weathering and natural bacterial breakdown. A slick hit early enough with enough dispersants is a slick that will not foul sensitive shorelines and marshes. The trade-off is this: The rapid injection of oil droplets into the water delivers a severe toxic shock to the top several meters of water for a period of perhaps six hours. In some areas, for example, an active spawning site or a productive marsh, the trade-off would be unacceptable; but in many areas, including substantial portions of the open sound, many scientists feel dispersants were worth the risk.

Lawrence Rawl, Exxon's chairman, has strongly accused the state of Alaska and the Coast Guard of preventing the timely and effective use of dispersants. He stated in the May 8 [1989] issue of *Fortune* magazine that "we could have kept up to 50 percent [5.5 million gallons] of the oil from ending up on the beach." That was, he later added, "if it worked perfectly." Steve Cowper, governor of Alaska, has in turn accused Rawl and Exxon of "a systematic and deliberate cover-up" because of such statements and has said Exxon was free to do all it could in dispersing the spill.

Both sides probably overreach. Rawl's arithmetic is at best wishful. It would have taken about 5,000 drums of dispersant to disperse half of the slick effectively. At the time of the spill [on Friday], Alyeska had about 70 drums on hand in Valdez and no plane to spray it with. Exxon's own numbers show that by Saturday the company had shipped in about 400 drums. By Sunday, when high winds began to make aerial activities hazardous and to spread the oil slick out of control, about 900 drums of dispersant were in Alaska. A week after the spill, according to the state, Exxon still had only a couple of thousand drums on hand.

It was sometime Saturday before two C-130 spray planes arrived, rigged with the proper nozzles and ready for action. Such planes and their pilots must be FAA-certified for the hazardous, low-level flying required, and very few of either exist on short notice anywhere in the world. Neither could the planes fly out of Valdez, because of the frequent low cloud ceilings that disrupt air traffic there. They had to be based in Anchorage, more than 100 air miles away. Had the two planes each been able to spray full loads of dispersants twice a day, they would have needed nearly half a month to treat half the spill—if things worked perfectly.

Spraying decisions in the first few days came down to one man, Coast Guard Commander Steve McCall, captain of the port of Valdez. Under a state-of-the-art plan agreed to earlier in the year by the federal government, the state of Alaska and the oil industry, McCall had the power to okay spraying immediately in large sections of Prince William Sound. McCall, however, was immediately thrust into what John Robinson, an NOAA scientist on the scene, describes as "an incredible situation, dealing with complex science—pro and con dispersants—with a governor, congressmen, corporate officials, panicked fishermen."

McCall, who called for test spraying of the dispersants Friday and Saturday before finally giving Exxon the go-ahead on Sunday, said he was strongly influenced by the concerns of the sound's large fishing community about dispersants' being more toxic than the oil spill itself. It is also widely accepted, he said, that on calm water such as existed that weekend, dispersants will not work—they need wave energy to mix them into the slick. He also described his thoughts when he went up in a chopper during the first test spraying on Saturday: "I looked at the size of the [slick], and I watched the plane drop its load [about 5,000 gallons], and it was like this." He dropped a pencil on the floor of his office, a room of about 250 square feet. "That load of dispersant on the size of that slick [was] about like a pencil on this floor.

"Did we hold Exxon up?" asked McCall. "Things were very disorganized. One day their plane sprayed dispersant on the *Exxon Valdez*, sliming a group of Coast Guardsmen. They never had the pieces of the puzzle together to do that much in the early days." Subsequently, state officials, citing Exxon's sloppy aim, also denied permission to spray more sensitive areas of the sound where McCall did not have sole authority.

Nonetheless, scientists like Robinson and David Kennedy, the principal NOAA oil-spill experts here, hold that what dispersants were available should have been used without hesitation nearly from the start. Both have been responding to the world's biggest spills for more than a decade. "I think once that ship hit the rocks, that was it, and will be the same again, given the current containment and recovery technology," said Robinson. Neither he nor Kennedy said he was convinced the dispersants would have helped that much, given the calm sea state, "but we were simply out of options, and [spraying] could not have hurt."

"You can always order more [dispersant] testing before giving permission to spray, and the tests are nearly always inconclusive—you usually can't tell for certain from the air how well they are working," said Robinson. And James R. Payne, who wrote the book on the subject, *Petroleum Spills in the Marine Environment: The Chemistry and Formation of Water-in-Oil Emulsions and Tarballs*, said, "Conditions from what I can tell were perfect for [dispersants], even with the calm sea state. They would have begun working to disperse the oil, and when a sea came up, they would have still been effective."

Behind all the heated political squabbling over dispersants, there may be some cold legal calculations. Both Exxon and the state of Alaska, which are inevitably headed for court on damages from the spill, are aware of a 1988 federal court ruling that significantly reduced claims from the 1978 *Amoco Cadiz* spill of 1.62 million barrels off the coast of France. The judge penalized the French government for restricting the use of dispersants. The French decision, said the judge, "seriously interfered with the success of [cleanup and] seems to have been solely the result of pressure from ecology and nature groups." Exxon has already charged in Anchorage Superior Court that Alaska hindered its use of dispersants and has asked for unspecified damages. The Exxon suit counters negligence charges filed by the state against the company.

The more one delves into the cleanup of a spill like this, the better prevention looks. But prevention had been eroding for years, the victim of both government's and industry's casual attitudes toward Alaska's environment, which have been exposed by the disaster. The examples that follow are not comprehensive, just a sampler.

In 1972: Alyeska and the U.S. Department of the Interior are struggling to overcome environmental objections to pipeline construction, and a lot of the concern is about the marine leg of the journey the oil will make. They promise state-of-the-art spill prevention, to include double bottoms on the tankers that will call at Valdez, but by 1977, when the pipeline opens, the Coast Guard has caved in to oil-industry resistance to the double-bottom requirement. Ironically, the yard that constructed the *Exxon Valdez*, National Steel and Shipbuilding, in San Diego, specializes in such construction.

In 1977: The state of Alaska asks Alyeska to have at least twelve miles of boom on hand to fight spills. The company insists that a quarter of this is plenty (approximately ninety miles of boom were brought in to Prince William Sound to try to contain the oil spill).

In 1978: The Coast Guard reduces the distance local pilots must stay with tankers after departing Valdez. No longer will they accompany ships as far as Bligh Reef.

In 1979–80: The state loses a suit with the oil industry over requiring industry to fund three positions to monitor Alyeska's operations. In the suit the state said it needed five full-time employees to carry out its legally mandated responsibilities. The legislature never funded the additional positions, and since 1980 less than one person's full time has been assigned to Alyeska.

In 1982: Alyeska gets rid of its only barge approved by the Coast Guard to hold oil from a major spill (5,000-barrel capacity). The barge must be taken every two years to Seattle for inspection to retain its approved status. It is replaced by a smaller, 3,000-barrel barge that is not Coast Guard–approved for oil-spill recovery. That same year,

Alyeska, to save money, disbands its full-time emergency-response team, dedicated to fighting oil spills and keeping equipment ready to respond. The company argues that by cross-training many other workers in oil-spill response, it will actually improve its capability.

In 1984: The Coast Guard's budget is cut, and its radar staff in Valdez, which monitors tanker traffic in the sound, is reduced from sixty to thirty-six. Its radar is also replaced with a less expensive model. At the time of the *Exxon Valdez* accident, the operators on duty were occupied with other duties and were not watching the screen, nor were they required to watch it.

In 1986: Exxon persuades the Coast Guard to let it reduce manpower by three persons on the *Exxon Valdez*. Two years later the Coast Guard admits in a letter to the company that it made a mistake in allowing the reductions, but since the ship has not encountered trouble, permission stands. Since the spill, evidence of mental and physical fatigue among the ship's crew has emerged as a theme in the NTSB hearings on the crash. Gregory Cousins, the third mate who was in command of the bridge when the ship struck Bligh Reef, had about seven hours of sleep (in snatches of four hours and three hours, several hours apart) in the twenty-four hours before the ship left port. The second mate, also exhausted, should have been on watch when the ship hit the reef, but Cousins had stayed on the bridge to let him get some more rest. "I am quite confident I would not be sitting here talking to you today if that second mate had been on the bridge," said Joseph LeBeau, an investigator for the Alaska Department of Environmental Conservation (DEC).

In 1988: Alyeska has been trying to get a new barge to hold oil from a spill, but the oil companies that own the pipeline have spent more than a year pondering the costs and seeking the best price. Finally obtained, the barge arrives in Seattle but cannot reach Valdez because of the difficulty of towing it through the Gulf of Alaska in winter.

In 1989: After a major spill in January (1,700 barrels in port, because of structural defects in the tanker *Thompson Pass*), Alyeska tries to expedite delivery of the barge but again must seek financial authorization from its owner companies. The barge does not make it to Valdez until a few days after the *Exxon Valdez* spill. Meanwhile, the barge that Alyeska is relying on to carry pollution-control equipment to the spill is in dry dock because the company cannot find a marine welder to fix it. Alyeska asks to use its equipment barge to hold oil recovered from the spreading slick but is denied permission because the Coast Guard fears it might be too unstable when loaded. The company has known for more than a year that the barge might not be approved in an emergency, according to Coast Guard officers. Tom Brennan, the Alyeska spokesman, declined to comment on the barge situation.

In 1989: Following the *Thompson Pass* spill in January, the *Anchorage Daily News* reports that twenty percent of the fleet engaged in the Alaskan-pipeline service is rated undependable. Another thirty percent of the fleet is rated very high. The *Thompson Pass* entered port with a suspected crack in her hull, but the Coast Guard and Alyeska decided to load her anyway. The company took the then unusual measure of "pre-booming" the ship during loading, which was fortunate, since she proceeded to leak crude oil. Of the eighty tankers that regularly call at Valdez, sixteen earned the lowest rating of the New York–based Tanker Advisory Center, the *Daily News* reported. Alaskan-pipeline tankers make up only thirteen percent of the American-flag ships of more than 10,000 tons but accounted for fifty-two percent of structural failures between 1984 and 1986. The problem, most maritime experts feel, is simply that the aging pipeline fleet sails some of the roughest waters in the world in the Gulf of Alaska.

In 1989: The *New York Times* reports that in 1988 the Department of the Interior suppressed warnings by the NOAA, the Environmental Protection Agency (EPA) and the Fish and Wildlife Service that said "current technology cannot effectively clean up a [major] spill."

Now, one might ask of many of these events, covering nearly two decades, where was the state of Alaska? The state, after all, had to sign off on Alyeska's contingency plan. And Dennis Kelso, commissioner of the Alaska Department of Environmental Conservation, has been talking tough in the national media recently about Big Oil's shortcomings. The DEC is the state agency responsible for monitoring the oil industry. Early on, Kelso became a minor folk hero around these parts by saying things that went right to the heart of what many Alaskans feel about the spill: "This is not a technical issue, not biological, not financial; this is fundamentally a moral issue. I don't think Exxon understood that what has changed, regardless of [environmental recovery], is Alaskans' sense of security, that promises be kept."

Kelso, a forty-two-year-old lawyer who moved here from Iowa and was a public defender in Inupiaq communities on the North Slope, seems sincere in what he says, as is the man who appointed him almost three years ago, Governor Steve Cowper. But both are lame ducks since the governor decided recently not to run again, and Kelso has been the exception, not the rule, in the history of the DEC.

In contrast to the high and ecologically correct profile he has presented the nation, Kelso's department is Alaska's smallest, next to Military and Veterans Affairs. In 1986 his predecessor agreed to Alyeska's demands that the DEC notify the company of routine inspections and that its inspectors be accompanied at all times by an Alyeska supervisor. This was not actually done, says the DEC now, but it is indicative of the regulatory agency's traditional toothlessness. "We're pleased to see [the DEC] standing up and fighting, but it's not been typical," said Dave Cline, regional vice president of the National Audubon Society in Alaska and a longtime resident of the state.

Cline, in testimony April 13 [1989] before Congress on the spill, described Alyeska's owner companies as having "actively lobbied" in the Alaska legislature to cut the budget of the DEC. That view is seconded by Gregg Erickson, Cowper's economist, and denied by Alyeska. For whatever reason, the DEC remains severely undermanned and under-funded by a legislature that only since the spill has begun to ask hard questions about oil-industry environmental practices.

Dan Lawn knows all about this, having been a DEC representative and Alyeska's sparring partner in Valdez for a decade (in the last few months he has been moved into another job at the DEC, reviewing oil-spill contingency plans). Lawn grew up in a pretty, forested part of Northern California where there were no jobs except logging, "and then they made most of the place a park. I was for the park, but I needed a job, so I came here as a contractor on the pipeline." He and a staff of three

were responsible for regulating drinking water, air quality, trash, hazardous waste and food-service establishments in a district that stretched
through more than 200 miles of coastal Alaska—also responsible for
monitoring Alyeska's terminal, one of the world's largest such operations. Lawn spent a long time trying unsuccessfully to get two or three
additional people just for the terminal.

Around 1983, Alyeska was allegedly told to cut costs by the oil
companies that own it—principally BP America (50.01 percent ownership) and Exxon and Arco (20 percent each). After 1983, Alyeska's
operating and administrative expenses declined sharply. In May 1984,
Lawn wrote a memo to his superiors that began: "Over the past several
months, there has taken place a general disembowelling of the Alyeska
Valdez Marine Terminal operational plan . . . severe personnel cuts . . .
plans and routine maintenance have been reduced drastically." He went
on to describe dozens of specific cutbacks, even in items like management's subscriptions to trade and technical publications. He also alleged
that oil-spill-recovery equipment was becoming outdated and concluded: "We can no longer ignore the routine monitoring of Alyeska
unless we do not care if a major catastrophic event occurs."

Through the years, Lawn wrote other memos, criticizing Alyeska's
responses to minor spills, noting cutbacks in operations, asking for help
in monitoring tankers calling at the port (it would take a dozen full-time
employees to inspect all the tankers, he said). Where did the memos go?
"They went into the system," he said.

"It's gotten no better since 1984," said Lawn. "They [Alyeska] wear
you down, they always have a reason, always a story why they didn't fix it
yet." Lawn is not universally popular at Alyeska. Tom Brennan once
called him "a kind of a jerk" in a news article (the Alyeska spokesman
later apologized for the statement to Lawn). Some employees at the
terminal, not particularly company loyalists, say they feel that Dan
Lawn has sometimes let his frustrations get the better of him, that he has
at times harassed people at the terminal needlessly.

"There are a lot of good, really good people working over there, but
we can't even say hi at the post office without risking them getting in
trouble," said Lawn. "I'll tell you how it is dealing with [management]

there. Once I asked them why they didn't have more people on a spill. I was told, 'That's a negative statement—be positive.' 'How?' I said. 'Well,' they said, 'ask us why did we choose to use the manpower distribution as we did,' or . . . I don't know, some shit like that. So I got written off as a jerk, a negative guy, because that's the way they think over there.

"Now, what I've feared for ten years has happened. You talk about negative attitude. If [Alyeska] had a positive attitude, maybe you'd have seen some response to the spill."

Disasters like this draw reporters from all over, but my eye was struck by the name of Jonathan Wills, which was entered on the sign-in sheet at a press center in Valdez as "Editor, the *Shetland Times*, Lerwick, Scotland." As it turned out, the spill was a natural story for him. Near Lerwick, some 30 miles south and 6,000 miles east of Valdez, in the Shetland Islands, is Sullom Voe, the giant marine terminal that ships a million barrels a day of North Sea crude.

Wills said I should compare the oil-pollution response and prevention on his side of the world with that in Valdez. He contended that Sullom Voe's operation, though certainly not perfect, was indeed close to what Alyeska's at Valdez had always been in word—state-of-the-art, world-class. This is all the more fascinating because the oil company that operates Sullom Voe is BP, whose American division is the majority owner of the pipeline and terminal at Valdez. BP America never responded to questions about Sullom Voe and about how much Alyeska's decisions are essentially BP's, but Alyeska's president at the time of the spill, George M. Nelson, was on loan from BP, as was Ivan Henman, also a key Alyeska executive. Another BP executive, James Hermiller, was appointed last summer to run day-to-day pipeline operations for Alyeska, and since October 1 [1989] he has been president. A fourth BP executive, Fred Garibaldi, is currently chairman of the owners' committee that sets Alyeska's annual budgets. Another oil company at both terminals is Exxon. It was a huge spill of crude oil eleven years ago by the *Esso Bernicia* at Sullom Voe that resulted in a revolutionary antipollution package there.

In a telephone interview not long after the March 24 spill, Captain Jim Anderson, deputy director of the Sullom Voe terminal, ran down a checklist, comparing how his terminal stacked up with the one at Valdez:

- Sullom Voe had random surveillance of tanker traffic by helicopter; Valdez had none.

- Sullom Voe had a round-the-clock pollution-response team of fifteen members, not including supervisors; Valdez no longer had a dedicated response team.

- Sullom Voe had booms on reels permanently deployed, ready to unroll to isolate sensitive environmental areas in the region; Valdez did not.

- Sullom Voe recently sprang a surprise oil-spill exercise on terminal workers, simulating a 35,000-barrel spill. It included bringing in additional skimmers, aircraft and dispersants. Valdez had never attempted a drill remotely that ambitious—or with no prior notification of top supervisors.

- Pilots at Sullom Voe ride with the ship past the last reef before open sea. At Valdez they do not go that far even with new, emergency precautions being taken since the spill.

- Sullom Voe had full radar coverage of all tanker traffic from the first reef encountered on the way to the terminal; Valdez did not come close to that.

- Four tugs assisted in berthing tankers at Sullom Voe. At Valdez it was usually two, sometimes three.

- All tankers at Sullom Voe were boarded and inspected by qualified marine officers. At Valdez it was less than twenty-five percent.

- Backup assistance for big spills (more than 17,000 barrels) is available to Sullom Voe within twenty-four hours. At Valdez that's unlikely, based on the recent spill. There has been no major spill since the changes at Sullom Voe. About 2,500 barrels total have been spilled there in the last decade in several smaller incidents.

The costs of upgrading pollution controls at Sullom Voe were paid by the oil industry, Anderson said. Another, perhaps less definable difference between his terminal and Valdez, he said, is that the local government controls all marine operations at the terminal, including berthing of ships. Industry controls only the land-based operations.

Ultimately, the difference between oil terminals may come down to something less tangible, call it a positive attitude if you will: "I was a tanker captain for eight years, and I've visited lots of ports [though not Valdez], and I'd say the major thing that sets Sullom Voe apart from many oil terminals in the world is not any special regulations but simply unswerving, strict adherence to them," Anderson said.

Valdez

SLOWLY, SPRING IS coming here. In a month they will be saying the place is as lovely as a Swiss Alps poster; a month ago they would have noted it is not that far from Siberia. These are the best of times and the worst of times for Valdez (pronounced Val-DEEZ), the little town that recovered from the devastation of the 1964 Alaska earthquake to win an All-American City award twice. Exxon is hiring literally thousands of employees for what promises to be a summer-long attack on the oil that by now has gummed more than 350 miles of shoreline out in the sound. Every restaurant, motel, bed-and-breakfast, and tourist shop is jam-packed. Money hasn't flowed here like this since pipeline-construction days, but the influx of reporters and photographers and oil workers has also bloated the population to several times its normal level. Squatter camps of people hoping to be hired have sprung up on gravel parking lots.

The sounds of float planes and helicopters ricochet constantly off the mountains that surround the town. Rent-a-car offices warn sternly against tracking oil into their cars. Banks, restaurants and filling stations sport help-wanted notices as employees quit to vie for the $16.69 an hour Exxon is paying for cleanup workers (and that's twelve-hour shifts, seven-day weeks, with time and a half overtime). Volunteer bird- and otter-rescue workers are organizing the First Bligh Reef 5K Run in an effort to avoid burnout. It may not get much local coverage because the editor of the *Valdez Vanguard* has quit to join the cleanup effort. A local fish dealer says he won't be able to make it this year [1989] even if the salmon run this May is not closed by the spill, because he can't compete with Exxon for workers. Local fishermen, their herring season closed due to oil, are hiring out their boats to Exxon for $5,000 a *day* for a fifty-footer, $3,000 for a thirty-footer, $800 for a skiff—"hush money," say some; "blood money," say those very few who refuse to take it. Damn good money, say most.

Valdezians have painfully mixed feelings about the oil industry these days. Their little town has an astronomical property tax base of $1.314 billion, of which Alyeska's marine terminal and pipeline account for ninety-four percent. It affords broad, well-maintained streets; a modern teen center, civic center and senior center; a fine library and museum; a town swimming pool; a paid fire department—in all there are eighty-nine full-time public employees for the town's 3,000 residents.

"We have a thousand more people than Cordova [the fishing port that is the other major town on Prince William Sound] and a budget that is ten times theirs," said Tom Gilson, the city treasurer. "In Valdez there isn't voluntarism, isn't much working with your neighbors. You work a week [of twelve-hour shifts] at Alyeska, you want your next week off. People in Valdez demand a lot of services and can afford to pay for them, and you won't find the cooperation you find in other small Alaskan communities until the last drop of oil flows through the pipeline.

"Who's to say we won't be better off then," said Gilson, whose grandfather came to Valdez in 1906 after the San Francisco earthquake. "Largely, the pipeline has been a blessing. I'm ambivalent about it, but I grew up in this town, and after I left school, I wanted to come back very badly, and I don't know, without all the pipeline development, whether I'd be living here right now."

Gilson talked about the town's anger and shock at the spill, then said, "In a way, it almost seems like we're not hurting enough. This is an economic boom like we haven't seen for a long time."

A month after its vessel cracked up on the rocks, Exxon is still taking a lambasting daily in the press—from the Coast Guard, from the state, from congressmen passing through—for what seems a botched and stalled cleanup effort. Debates rage over whether to flush the beaches with cold water or hot water—or whether to steam-clean them. Meanwhile, every day the *Glacier Queen II*, a tourist boat, sails from Valdez at 6:00 a.m., loaded with beach cleaners. It takes them four hours to reach the nearest devastated beach and four hours to get back. The four hours in between they spend, slipping and sliding over the gravel and cobble, wiping individual rocks with paper towels and stuffing the soiled tissues into big trash bags.

Exxon's public-affairs people often seem whipped and don't even bother to respond to the many allegations of Exxon incompetence that routinely circulate. If you believed them all, you would wonder how the company could run a filling station, let alone a giant multinational business. If you talk to the oil-spill experts, both those hired by Exxon and the state and those from the NOAA, you get a different picture. The situation here, they will tell you, is chaotic, frustrating, controversial— and just about what you get in every big oil spill they have seen. These veterans all agree that, months from now, when all the shouting is done, Exxon will likely have recovered something like ten or fifteen percent of the 11 million gallons it spilled. That is the depressing but standard recovery seen in most big spills of this nature. Twenty-five percent would be tops, the experts say. The contingency plan Alyeska is so proud of promised fifty percent. Alyeska's president, in a memo to his employees, has offered this advice for trying times, which he said he got from a well-wisher: "Take two aspirin and try to get some sleep." A locally popular T-shirt, picturing an otter with a tear trickling down its furry cheek, has some better advice: "One ounce of prevention is worth 11 million gallons of oil."

Exxon does get high marks for something. It got here fast with virtually unlimited financial resources, and it did not quibble about writing checks in an effort to get the cleanup going. Ironically, the huge corporation's almost unimaginable wealth has created something of a backlash. People here who have read how even a $1 billion cleanup bill will scarcely dent Exxon's balance sheet feel that somehow it should be harder for the company. Exxon should hurt more, realize more deeply that money can't solve all the problems the spill has created. Several fishermen told me of the elderly native woman who gave a bouquet of flowers to Frank Iarossi, Exxon's shipping-division president, at a hearing on the spill held in the native village of Tatitlek near here. He later remarked, so the story goes, how deeply he appreciated that show of support at a difficult time, never realizing that the woman's gesture meant she hoped to see him at his funeral. "Exxon can only conceive of saying to people, 'Show us what you've lost and we'll pay you,' " said Mike Lewis, media director for the Prince William Sound Community College and a member of the environmental group Earth First! "The people in Tatitlek are being forced from the subsistence into the money economy with Exxon largess, and some may never go back."

Time and again, when I have gotten numb trying to sort claim from counterclaim about responsibility for the spill and its cleanup, I go visit the animals, the birds and otters, oiled by the spill, brought from all over the sound to rehab centers hastily erected here. They are the only true innocents in this whole, sad business. They are not consumers of petrochemical products, nor sellers of salmon. They are neither regulators nor environmentalists, are not in any way responsible for what happened. They just are, and now many of them aren't anymore.

Over at the bird center, auburn ponytail flipping as she moves swiftly about business, Jessica Porter is tubing a murre. Each of the several dozen birds recovering in the converted classrooms of the Prince William Sound Community College gets tube-fed three times a day. Murres, pretty, little waterfowl that appear to fly underwater when diving for food, took a heavy hit from the oil. Hundreds will be recovered, thousands likely died and were not recovered, says Porter, one of the veterinarians here. (The total estimate of birds killed by oil will eventually go over 100,000.) Another murre, which arrived several hours ago looking like a sullen tar ball, has been resuscitated and cleaned with a one-percent solution of Dawn in warm water. Dawn, says Porter, after years of experimentation, is still the only product that cuts the grease and is still gentle. "We get lots of ambulance chasers at these spills, trying to sell us all sorts of products, but nothing beats Dawn," she says.

Soaping the murre has taken two volunteers working with toothbrushes and Water Piks nearly an hour and fifteen tubs of soapy water. Next comes another session of forty minutes, rinsing the murre with high-pressure hoses until, improbably, its feathers become dry. "It sounds crazy," explains Porter, "but what the oil does is disrupt the integrity of the feathers, which form a protective basket around the bird, trapping air. As the oil is removed, the feathers actually become so dry you can blow in them as you rinse." The murre, pinioned by beak and wing and one webbed foot, is pissed but fluffing nicely now under the water jets. "Hang on, murre, hang on!" says Jim Noland, one of the washers, but no, "back off, he's had it for the day . . . See his eyelids coming over his eyes; his heartbeat's up, too." They will try again tomorrow with the murre. The birds coming in now, says Porter, are goopier but healthier than in the early days of the spill, as the more toxic elements of the crude—the light ends, like toluene and xylene—evaporate, leaving the heavier asphaltenes, waxes and paraffins.

Porter, raised at Tenth Avenue and Fifty-second Street in New York City, worked her first spill in 1964, in the marshes of East Anglia, at age fifteen. "It was horrible—we were using acetone, a solvent, for God's sake," she says. "It cut the oil, but the inhalation was toxic to both the birds and the rescuers." Nowadays she rehabs hummingbirds, otters, foxes, raccoons, muskrats, deer, elephant seals and other wildlife at her Wolf Hollow Wildlife Rehabilitation Center in the San Juan Islands of Washington State. "It doesn't pay like doctoring poodles in Honolulu, but it's a living," she says.

Bird rescue, says Porter, has always been the orphan of oil-spill cleanup, "getting the most PR and the least money. Originally, it was unscientific, and it got this image of little old ladies whose kids left them with an overabundance of maternal instinct and lots of spare time. But the people running it these days are professional zoologists, biologists, vets. Wildlife rehabilitating is becoming a science." Her employer here, the International Bird Rescue and Research Center (IBRRC), was contacted by Alyeska, for which Alice Berkner, the founder of the IBRRC, did training sessions two years ago. "What Alice has done is standardize techniques for cleaning birds and making sure they are ready for release that can be applied anywhere in the world," says Porter.

I asked Porter how she justified the cost, estimated at more than a quarter million dollars to date, of paying boats to range the huge sound, chasing down a few hundred birds, many of whom will die, all of whom together, whether they live or die, will not affect the ecological balance of the region. Her answer was good enough to serve for all future spills: "Because they are citizens of this planet, too, and we are responsible for them . . . their right to do what they do in a healthy, clean, free environment. I think more people are recognizing these spills as a crime against creatures with no redress in Congress or the courts. This is just taking responsibility for what we did, all of us who drive cars and use plastic cups."

In the background as we talked, a volunteer was busy converting shot glasses to cc's. She figured it would help fishermen who were picking up the birds under contract to Exxon calculate how much fluid was needed to hydrate the victims on the way into the center. Another volunteer supervised a day-care center for the children of rescue workers. The bird rescue center, for all its makeshift surroundings, smacked of organization, training and single-minded devotion to its purpose rarely seen in other sectors of the spill response.

"IBRRC was called two or three hours after the spill," said Porter. "I was contacted at 6:30 a.m. Friday [March 24, 1989] and given half an hour to catch a plane. Four of us were in Valdez setting up operations on Saturday morning. I think we need the same kind of SWAT-team type of response to the spills themselves."

Perhaps it would work. But why is it I can just see Alyeska and Exxon, if they ran the bird operation, rejecting volunteers just as they rejected help from fishermen on the spill; citing the danger of bird bites, risks of infection, corporate liability, inadequacy of housing? . . . What I think is really the secret of the bird people's getting up here and bending to work so wholeheartedly is not just organization, or experience—it is because they care so damn much.

From the bird rescue center, it's a short walk between dirty, bulldozed banks of old snow to the old Growden-Harrison Elementary School, which more resembles a keening, squalling MASH unit these days. It is home to some seventy or eighty otters just now in all stages of recovery and disintegration from the dark kiss of the *Exxon Valdez*. Early on, things were a lot worse, says Terrie Williams, a marine-mammal expert up here from the Sea World Research Institute, in San Diego. One large otter came in, bleeding and duct-taped from head to toe because the fishermen who found it had nothing to restrain it with. It died soon after. The rescuers knew that oil would disrupt the otters' insulating fur, and the mammals, which have no insulating layer of subcutaneous fat, as do seals, would die in the icy water that is otherwise their natural home.

But that didn't prepare them, say workers, for what they got—otters rattling, and gasping, contorting, foaming at the mouth, excreting blood. Autopsies revealed that the real damage often was internal— livers that crumbled to the touch, lungs blown like broccoli, immune systems defused, all from those light ends, the toxics coming off the crude in the initial days of the spill. The otters inhaled the fumes, absorbed them through their skin. The pain they suffered, says a vet, "must have been indescribable."

"We know a lot more now than we did about the logistics of trying to save animals in a big spill," says Terrie Williams. "We know now it's a nightmare."

Things at this point are more settled. There have been no deaths for days among the inmates, which are housed in pens under the score clock and backboards of the gym. A sign on the wall reads: "All six otters that went to Tacoma are doing fine. . . . They had a good trip and are all eating. Yea!" Many of the otters have been named—Fat Albert, Ollie, Otteri, Garfield, Odie, Kimmer; one, fur slicked heavily with crude, has been dubbed John Tower. Exxon has given the recovery center here "a blank check," says the vets.

"The critical thing," says Williams, "is to get them grooming their fur again to get the air layer back into it, get their natural oils flowing. If they don't groom, they die of exposure." As of late April the center has handled about 140 otters, of which about half have died or been euthanized. Long-term survival is still questionable for many of the living. Estimates of otter deaths in the sound are running as high as 6,400— perhaps half of the local population. Some commercial fishermen in Cordova, when they first heard of the otters' plight in a meeting with Exxon, stood and applauded. They are convinced that the expanding otter population (otters are federally protected from hunting or trapping) is adversely affecting their catch. Others recognize that nothing, including their own endangered livelihoods, has so mobilized public sentiment about the tragic nature of the spill as the travail of oiled otters on the world's television screens. The crush of media at the recovery center in the early days of the spill probably killed some otters or hastened their demise, workers here acknowledge.

The last otter I saw in Valdez was a pregnant female that had come in days earlier with corneas scarred shut from the fumes. She had been moved to an outdoor cage, a good sign, but now she has just aborted her pup. The female lies listlessly on the floor of the plywood and wire cage, as another otter licks and grooms about her sightless eyes and—there is no other word for it—cuddles her. I know the journalistic pitfalls of the Bambi syndrome, of attributing human emotions to dumb animals, but I also know that the great ethologist Konrad Lorenz, after a lifetime of scientific study of animal behavior, felt strongly that "in terms of emotions, animals are much more akin to us than is generally assumed." I haven't the least trouble believing that.

I think it would be fitting to invite some of the executives of Alyeska and its owner companies, and maybe a few tanker captains who haven't yet got as careless as Joe Hazelwood—invite them to spend some time as volunteer otter handlers. The men who can ram one of the world's biggest pipelines across untrammeled Alaska and navigate superships through the planet's stormiest ocean then could try something really challenging, like putting the natural oil back in a wild creature's fur, or figuring out what to do with a blinded otter.

Exxon just keeps asking for punishment, pumping out press releases daily that start like this: "Exxon is ahead of schedule on the shoreline cleanup" (April 24 [1989]); or "Deployment of shoreline crews to clean Prince William Sound is proceeding more rapidly than planned" (April 26 [1989]); or "Exxon's shoreline clean up operations are accelerating" (April 23 [1989]). Some days, it seems, it might be better not to write anything.

Take the day that one Thomas Copeland, a fishing captain out of

Cordova, rolled into Valdez Harbor aboard his seiner, the *Janice N.*, with 1,500 gallons of crude oil aboard. Copeland is eating Italian carryout, standing over a fish tote containing about 50 gallons of oil, entertaining the local press with his opinions on things like Exxon's beach-cleanup crews, which he describes as a "buncha piss-tested rock wipers" (Exxon contracts carry a no-drug-use clause).

Say this about Copeland, he does not discriminate. Of some of his fishing colleagues in Cordova that I am headed down to interview, he says, "Hell, only the pissers and moaners and welfare cheats are left down there. If you got any balls, or self-respect, you're out there on the spill." No one, least of all Exxon, he says, "really wants to attack that oil, to get it back up."

Copeland seems to speak with some authority. The next day the Anchorage papers will report that Exxon, with forty-seven skimmer vessels spread around the sound, captured only about thirty percent more oil that day than the *Janice N*. And how did Copeland do it? He did it with Denny and Christine, two crew members who sat on the stern and dipped the oil up with scoops and five-gallon buckets as the captain backed into tide rips and coves where currents had piled up the crude into thick ridges—"like digging into calves' liver," says Christine.

Copeland paints a picture, generally corroborated by people who have been out to the spill, of skimmers that don't skim, of those that do waiting all day for a barge to unload their recovered oil, of pumps that burn up trying to pump up the thick crude and of high-tech cleanup vessels that simply cannot work if there are waves on the sound. Even a mammoth 425-foot skimmer sent by the Soviet Union finally went home in defeat after the oil became so thickened and mixed with seaweed it choked the big Russian's pumps. The oil Copeland scooped up he sold for five dollars a gallon to Exxon, he says. He recalled that early on in the spill, Exxon rejected suggestions from fishermen that the company simply put a bounty on the spilled oil and let everyone with a boat and a will have at it.

Copeland has come in from the spill only to try and upgrade his operation by mounting a huge septic-tank truck on another vessel to suck the oil he corrals into a 112,000-gallon agricultural silage bag.

When that deal falls through, the *Janice N.* sets off into the sound in a hard, cold rain, with dark coming on, buckets piled high on her afterdeck.

Ever since that encounter, the thought of Tom Copeland has made me think of the Alyeska executive who explained why the company's initial response to the spill took so many hours in coming. I remember his saying, yes, in the middle of North America's biggest oil spill they did send the night-shift workers home when their time was up, "because it's not our policy to burn people out." Probably, when you are running a big company, that makes sense; possibly, it didn't make much difference when Alyeska was responding to a spill that massive. But you have got to wonder how it would have gone that night if you had had Tom Copeland in charge.

The Battle of Sawmill Bay

FROM A DE HAVILLAND Otter floatplane, high over the sparkling waters of Prince William Sound, the beaches are gray-black as far as the eye can see. They always have been. It is the natural color of the glacial rock, and from a distance it is sometimes difficult to tell the hundreds of miles of oiled shoreline from the thousands of miles that are still pristine. The sound today [spring 1989] is majestic, a royal canvas for the morning light that daubs the snow peaks, silvers the dozens of waterfalls that slice down rock and evergreen slopes to feed the translucent blue-green waters of the fiords. The scale here is one of geologic immensity, a land where the biggest living things are glaciers, growling and booming in retreat from an ice age that ended for the rest of the earth millennia ago. As we pass Bligh Reef, the water is freaked with small icebergs, calved from the Columbia Glacier to our right. One is an odd little zebra, its bluish white ice striped with crude.

South of the *Exxon Valdez*, which has been towed to Naked Island and moored for hull repairs, we begin to pick up sheens of oil. Not a boom or a skimmer is in sight. Just beneath the surface, I spot what looks like a sizable pool of dark crude oil or perhaps a reef; but no, says the pilot, that is a mammoth shoal of "ghostfish," herring just now entering the sound to spawn. The herring season has already been closed because of fears oil would contaminate the catch. We bank low over a shoreline where the herring are spawning in the sunny, gravel-bottomed shallows, discharging roe and milt, turning the water milky for hundreds of yards with the procreative fury of spring. The same scene is repeated on the next island south, but here a rainbow sheen of crude oil swirls with the creamy herring spawn in the lemony green waters of the near shore. It is an unholy brew, as lovely a picture of life meeting death as ever I hope to see.

We are hitting serious oil now, still moving through the sound, tarring shorelines after all these weeks. It is these edges you worry about, more than the open water. So many of the fish, especially in the earlier, more vulnerable stages of life, prefer to hug the shore's edge for its calm water and protection; similarly, a great deal of the bird life is concentrated on the land-water intersection, nesting and feeding there. Even the wide-ranging open-water species, like seals and sea lions, must come to the edge of land soon to have their pups. Land dwellers like the edge, too. Exxon back in Valdez has large charts of the sound detailing "areas of intensive Black and Brown bear use of the shoreline," one of the many factors in this wilderness that makes cleanup difficult. There is no thought, with hungry bruins waking after a long winter, of encamping workers on the beaches overnight.

The shoreline cleanup operations I can see after two hours aloft appear so insignificant against the wide compass of the sound, like sticking straws into a lake to try and suck it up. David Kennedy of the NOAA has cautioned me that for all the beating Exxon is taking on its slow-paced cleanup, there are worse things than not immediately trying to clean every mile of shoreline.

"The problem is, you really need to wait until all the oil has moved through an area before you try to clean the beaches," said David Kennedy. "It's going to be an eyesore—it can gunk up birds or marine animals—but you will do much more damage if you send crews out there six or seven times to clean it every time more oil hits. For example, Exxon had one cleanup demonstration with thirty to forty cleanup workers and about fifty press, and they all tracked through the oil and worked it way down in those sediments to where no one will ever get it out now. In the *Amoco Cadiz* spill they did more damage to the marshes they cleaned than to the ones they left. There are exceptions, such as where you've got rookeries, herring spawning and marine mammals pupping, or where the oil is pooled so deeply it keeps reentering the water on every high tide. But for the most part, once you've nailed a beach, it's nailed, and rushing out there isn't going to do much."

This is doubtless sound advice, just as it's true that for every mile of beach that is oil-stained, there are many, many miles that aren't. It's not that the statistics lie. They just don't correspond to what you feel. Seeing an oil slick floating among these grand canyons of balsam-smelling forests and clear waters is a bit like a guest's coming up and telling you he has just crapped in your swimming pool, but don't worry, because the fecal matter floating out there only covers one-tenth of one percent of the water surface.

More than an hour by air out of Valdez, eight hours from anywhere by boat, we set down in one of the loveliest alpine scenes on earth—the Rocky Mountains with Lake Geneva dropped into the heart of them. It is Sawmill Bay, the sole province of about sixty people, mostly Aleuts, who live in the village of New Chenega, and about a hundred million salmon. If there is a single pocket of Prince William Sound where the oil could destroy things that are nearly irreplaceable, it is here.

The Chenegans say their people have endured for thousands of years in these parts, hunting, fishing, living off the land then as now. But perhaps never have they been tested so sorely as in the last quarter century. The Good Friday oil spill that now threatens their subsistence lifestyle came twenty-five years to the day after the massive earthquake of 1964, which rearranged thousands of square miles of Alaska as easily as one might rumple a bed sheet, destroying their old village fifteen miles north of here. Twenty-three of the sixty-eight closely related inhabitants of old Chenega were swept away by the tidal wave from the quake.

"It's so sad to see this, because the trauma of 'sixty-four has never really left the people—you grieve a long time for loved ones whose bodies never were found," says Gail Evanoff, whose husband's parents were among the missing. She remembers living afterward in Cordova, one of the places to which the surviving Chenegans scattered: "Larry and I would listen to the elders talk, hear the sadness at the loss of community, of the culture disintegrating . . . the longing to go home."

Other communities around the sound, like Valdez, she recalls, got quick and generous assistance to rebuild, while Chenega got nothing. The long process of founding New Chenega would probably not have begun without the pipeline. Frantic to begin construction of the pipeline in the early seventies, Alaska finally acceded to federal demands that it first settle native land claims throughout the state. More than 40 million acres were turned over to Aleuts and other native peoples. After years more of lobbying dozens of government agencies, a site on Evans Island was chosen for New Chenega, on land owned by the native Chugach Alaska Corporation. Houses, diesel generators for electric power, gravel streets and even streetlights were erected in the wilderness. The resettled community was just entering its third phase, developing a fisheries industry, when the spill threatened.

Now it is the uncertainty that is so hard to bear, say Frank Gurske and his wife, Sue, as we talk in their home. From the window of their modest modular house is a view of blue water and snow peaks that would bring millions at Vail or Aspen. "We pretty much rely on halibut, deer, sea mammals, black bear and salmon for our food supply . . . now the state Division of Subsistence has advised us not to hunt, not to fish," says Frank. "That puts the pressure on, you know? Can't hunt, can't trap, can't fish . . . I feel like we're done for. Hey, you know what one of the worst things is . . . kids, they run so free here, and now they can't even play down by the beaches [lest oil wash ashore]. Mine are getting real bored, fast."

Exxon's financial response—the part it has seemed best at throughout the spill—has reached even New Chenega. Here and in Tatitlek, the other native settlement on the sound, every adult has been put on beach-cleanup wages—$16.69 an hour, even though few of the natives have actually been called by the company to do any work. Every week or two an Exxon supply ship arrives bearing yogurt, Granny Smith apples, several varieties of bread, an assortment of meat and poultry, canned goods. "Those Granny Smiths are good eating, but you've got to feel like Exxon's buttering us up," says a neighbor of Frank Gurske's.

Exxon will even pay for mental-health counseling, says John Totemoff, the fifty-nine-year-old president of the village council, "but people here thought they'd rather have a priest come instead." Alaska's Aleuts, who often intermarried with their Russian colonizers in the eighteenth and nineteenth centuries, are predominantly members of the Eastern Orthodox Church. This morning much of the village is on its way down to the little wooden church, set near a rushing brook. "Lord, deliver these thy servants from calamities and destruction. . . ." Father Simeon Oskoloff, a Russian Orthodox archpriest flown down from Anchorage, swings his brass censer vigorously, spreading the sweet odor of incense through the building, chanting, singsong, now in Russian, now in English. Sun slides off the hemlocks, glints from the higher slopes, filters through the windows of the airy, whitewashed structure. Outside the door a child pees off the back steps. On the bay salmon boats pull oil-containment booms back and forth as the priest inveighs against the spill, sprinkling the congregation with "a special holy oil that was brought from a monastery in Russia." Holy oil thrown against Exxon oil. It seems at least worth a try. "They are pretty adamant about staying here," says Oskoloff after the service. "They have just come home after a long time away."

From the air, this is Boom Town, U.S.A. The spill response has produced its own art, a kind of calligraphy of oil cleanup, "boom art," if you will. We have sorbent boom, harbor boom and sea boom; ocean boom and Goodyear boom and Norwegian boom; solid boom and inflatable boom and homemade boom—the last made of spruce logs chained together. There is orange and gold boom, ivory boom, black and red and gray boom. Boom encircles little islands, like a work by Christo; boom necklaces the mouths of coves and inlets and collects like plates of Technicolor spaghetti where eddies have formed.

The reason for all the boom, which acts as a floating fence, with "skirts" that extend below the surface to trap oil, is found at the head of the bay, across from New Chenega. There, in a series of green-roofed buildings reminiscent of a Bavarian chalet, is one of the largest salmon hatcheries in the world, the Alaska fishermen's equivalent of Alyeska's pipeline. Since last fall, 126 million baby pink salmon have been incubating in endless corridors of stacked stainless-steel trays, designed to simulate the gravel-bottomed streams of their natural habitat. Now, just as oil is poking tentative fingers into the mouth of the bay here, the inch-long salmon fry are at the stage that they must be released into outdoor holding pens; within days, when the plankton they feed on begin to bloom in Prince William Sound, the salmon must be set on their two-year journey through the Pacific Ocean. How the survivors (about 7 million fat, three-and-a-half-pound fish) find their way thousands of miles out and back to this precise spot is a secret held in the little stream, no wider than a roadside drainage ditch, that splashes into the bay almost unnoticed beside the impressive structure of the hatchery.

A classic experiment done years ago captured spawning salmon as they returned to their home stream and hooked them to an electroencephalograph. Their brains were almost shut down, all bodily functions bent on procreating, except for the brain centers that controlled the sense of smell. When water from the home stream was passed through the captive salmon's tank, their olfactory bulbs began firing madly, sending spikes leaping across the recording paper of the brain-wave machine. The natal water was diluted by half with distilled water, and again the same response. It did not disappear until the diluted water had again been diluted by half, and so on, for the tenth time.

All our science cannot explain this. "We think each home stream acquires from the soils and plant communities of its drainage basin and in its bed a unique organic quality which young salmon learn in the first few weeks of life," writes Arthur D. Hasler in *Underwater Guideposts*, an examination of how salmon home. Marcel Proust, who was not writing about salmon, nonetheless captured the phenomenon exquisitely in *Remembrance of Things Past*: "The smell and taste of things remain poised a long time, like souls, ready to remind us, waiting and hoping for their moment . . . and bear unfaltering, in the tiny and almost impalpable drop of their essence, the vast structure of recollection."

Or as Armin Koernig put it: "The salmon is an animal destined to be ranched." Koernig—an emigrant from Germany who "believed all those get-rich fishing stories" and moved to Prince William Sound in 1963— pioneered the concept of culturing salmon here, letting them range the oceans and grow fat, then harvesting them when they unerringly returned to their birthplace years later. Natural salmon reproduction in the sound was historically world-class, but after the 1964 earthquake rearranged the topography, effectively lowering water levels in many spawning streams, there were more bad years and fewer good years for the wild salmon returns, say fishermen. Combined with overfishing and poor fisheries management, this resulted in closed fishing seasons in 1972 and 1974. About this time the fishermen had just lost their battle to stop the oil pipeline from coming to Valdez, along with the marine terminal and the tanker traffic through their fishing grounds. "Either we did something about our future or we went to work for the oil companies," said Koernig.

As it has inevitably turned out with so many aspects of Alaska's development in the pipeline era, "the key to starting up the hatcheries was the oil revenues that began flowing to the state," said Koernig. With the newly anticipated oil money, the state formed a division of fish production and floated loans of about $9 million for development and start up of the hatchery. Later a second, even larger, hatchery was added at Esther, on another part of the sound. In the last few years revenues from the five percent or so of released salmon that make it back from the ocean have totaled more than $40 million for the two hatcheries, which each have annual operating costs of around a million dollars a year. Salmon fishing in the sound now depends at least as much on hatchery fish as on the natural returns to thousands of streams and other spawning areas. No one is sure how much expansion is possible from the current releases of more than half a billion fry; it depends on the ultimate carrying capacity of the "ranch"—a sizable portion of the North Pacific. Some scientists think releases of a billion are easily possible, and 2 billion fry are not inconceivable.

"And this is a renewable resource, a system that is designed to go on and on, long after the oil has dried up and the pipeline pumped its last—

if we just keep the system healthy," said Koernig. And while nothing produces instant revenue like oil, he and others note that frequently the price of the most desirable varieties of salmon, like sockeyes, or "reds," has substantially exceeded the value of a barrel of crude by several dollars apiece ($21 to $12 in 1988). And if Alyeska is producing an eighth of our nation's oil, "well, Prince William Sound this year probably will produce a pound of top-quality seafood for every American," said Koernig. There are pitfalls in salmon aquaculture—disease and genetic problems associated with inbreeding—but the Alaskan system so far has avoided these. To the extent Alaska can recoup its sullied dreams of a wild environment coexisting with a booming economy, of having its cake and eating it too, perhaps it is with the salmon that that hope leaps highest.

On Sawmill Bay, Eric Prestegard has had scant time for such philosophizing in recent weeks. He manages the big hatchery for "Pizwak," the nonprofit Prince William Sound Aquaculture Corporation that operates the hatcheries for the benefit of the sound's fishermen. Only on this day, more than a month after the spill, is he directing the booming of the bay with several thousand feet of mammoth Norwegian sea boom that just arrived on special order, along with its own Norwegian to advise on its deployment. The boom is superior to virtually anything obtainable in the United States for protecting the bay in rough weather. In one of the thousands of little glitches that plagued the spill effort, Exxon spent several days trying to call the Norwegian supplier but didn't know it had moved half a block and changed its telephone and fax numbers. Incidents like that have been the norm, rather than the exception, said Prestegard, in what came to be termed the Battle of Sawmill Bay. Had it not been for rapid action by the fishing community, he and many other observers of the spill feel, the big hatchery would have been disastrously oiled before Alyeska or Exxon reacted. Perhaps, had they survived, the young salmon would have been imprinted with the smell of Alaskan crude oil and spent their lives following a supertanker around, went one local joke.

"We already had a half million fry in the water [in holding pens] when we got the call the oil was coming our way," said Prestegard. "Five days later we still were in a state of chaos here, getting some boom dropped from helicopters, but often not the right kinds, or not enough to do any good. The [oil industry] kept going on the media saying, 'Boom is deployed.'. . . Well, there's a big difference between dropping boom all over the sound and actually getting the right boom, getting it anchored so it won't break loose, getting the type that will withstand waves . . . it was just chaos."

In Cordova the morning of the spill, recalled Paula Lamb, she thought "someone had died," from her husband's tone of voice on the telephone at seven o'clock. "Then I heard him converting barrels to gallons," she said. "That was the most helpless time of all, those first few days, for all the fishermen. They are men who make decisions—not always the right ones, but they don't sit around. They were saying, 'Give us something to do,' and we could have had fifty boats out there that morning, and Alyeska never even returned our telephone calls or said, 'We'll get back to you,' and this went on for nearly three days."

On Monday, Paula's husband, Jack Lamb, a vice-president of the Cordova District Fishermen United, made some calculations about how the tides and currents would distribute the slick. "And a guy with the state DEC said, 'Hey, you are the first people we've talked to that know anything about Prince William Sound,' " said David Grimes, a herring fisherman out of Cordova. Monday night, Grimes continued, the state managed to get him, Lamb and a few other Cordova fishermen into a top-level planning meeting with the Coast Guard and Exxon. "I felt immediately there was an incredible amount of energy going into trying to look good—a power struggle between the state, which wanted any action, and Exxon and the Coast Guard, both very worried about the propaganda ramifications of whatever they did," said Grimes.

The fishermen showed the assembled officials where the oil would be carried by the currents and pressed the critical need to boom the hatcheries. "We really felt [Frank] Iarossi [Exxon's top man on the scene then] perked up, like he was getting some information he could use for the first time," said Grimes. Iarossi, according to state officials, immediately gave the fishermen a virtual blank check to procure boom and other supplies, reportedly writing on his business card that Exxon was good for a million dollars and more if needed. "Iarossi could make decisions, and I really think he was trying to do his best, but he seemed under the handicap of a terrible bureaucracy," recalled Grimes.

On Monday night the fishermen and state officials, racking their brains about how to save Sawmill Bay from oil, hit on the scheme of importing the "supersuckers," the giant vacuum cleaners made specially for the quick recovery of oil spills on the North Slope drilling grounds. Exxon agreed and brought them down the 800-mile service road built for the pipeline.

"Exxon said they did everything the fishermen asked them to do," said Jack Lamb. "The real problem is that's all they did . . . We are fishermen, not oil-spill-recovery experts. Exxon never took the initiative, and as for Alyeska, they simply faded from sight—I hold them totally responsible." Unwittingly echoing the old Alyeska motto of pipeline-construction days, David Grimes said, "One of our advantages, I think, was that we didn't have anyone telling us we couldn't do something . . . Exxon always wanted to know 'who authorized that . . .' "

To Paula Lamb, there was another tragedy that was part of the mess Exxon had made of her sound: "People in Cordova are pretty patriotic, and I shudder to think what the world media is going to do to the U.S. We criticize the Russians on the quake they had in Armenia, and now look at us, the richest capitalist country in the world, and we didn't reinvest enough profits to protect an environment like this. It breaks my heart. I wouldn't want to live anywhere else, but we're all out of control in this country. We have to give up some of the largess from oil to bring the Exxons and Alyeskas into control."

Back at the hatchery, Eric Prestegard was unwinding a bit after nearly a month of solid eighteen-to-nineteen-hour days and reflecting on the future of his hatchery. Within days he would release the millions of fry, for good or ill, on their two-year journey. "My fear is, with the oil on the beaches, that the fry always hug the edges of the sound for several weeks until they head to sea," he said. Once he releases them, like a shepherd still tending to his flock, he always canoes along the beautiful indented fringes of nearby Knight Island, watching the little silver and black salmon fin along its shallows, feeding and resting. "It's so beautiful, my favorite place in the world—at sunrise and sunset there's a thousand little bays and inlets catching the light," he said. "But that slick came down both sides of Knight Island, and it's gone."

Cordova: "Our Way of Life Is Threatened and Nobody Seems to Give a Damn"

SOMEWHERE IN THE forty miles of Alaskan coast that separates Valdez from Cordova there is an undefined but palpable break point, where the thirty feet of snow a year in the former turns into the 200 inches of rain a year in the latter; where the stark grandeur of the landscape untenses a bit, the mountains are gentled in their plunge to the sea by alpine meadows and bogs that, come summer, will blaze forth wildflowers. In the waters, loons glide, orcas lunge and roll, and blithe otters float on their backs, peering curiously at our passage down a coastline cloaked in hyperborean rain forests. Through the mists, it seems like we are rediscovering America, but it is only the regular ferry run from Valdez to Cordova.

Valdez, with its postquake grids of neatly planned, treeless streets and modular housing, hints of artificiality, of all-plopped-down-at-once: stores over here, houses there, public buildings in this sector. Cordova, rambling up the ledgy slopes of Mount Eyak, is organic; Laura's Liquors is next to the Faith Lutheran Church, which is next to a restaurant, and so on. From every alley and back yard and garage peek fishing boats, and piled around are enough seine and gill nets to enmesh the earth. The town, with many buildings left over from the copper-mining days of the 1920s, has a ramshackle appearance, but its boat harbor is strictly top-of-the-line, from the well-maintained breakwaters and floating docks to the modern Fiberglas and aluminum boats with jet drives, big Caterpillar diesels and state-of-the-art electronics for navigation and fish finding. In the summers, when fishing and the canneries are at full blast, the population swells from 2,000 to 6,000.

Cordova and Cordovans run on energies that are more natural, pulsier, peakier than the piped oil that is Valdez's lifeblood. Life here is tied to the caprice of herring and salmon and halibut, of dungeness, tanner and

Alaskan king crabs. Fishing season for a given species may open on a few hours' notice, then may be over within a few hours more at the whim of the state biologists who determine how many fish can be harvested and still maintain a healthy, long-term population. One hang of the net on a bottom obstruction at a critical time, or an engine that falters untimely, may be the difference between fortune and failure. For two weeks in Valdez, I heard few people speak of "adventure" and "excitement." In Cordova, I almost never heard anyone not mention those as a reason for being there.

There is fierce competition on the water here but camaraderie on the land, lots of neighborly potluck suppers, an island sense of community— the town's fishermen have so far resisted business and tourism interests' attempts to connect Cordova with the rest of Alaska by road. "Plenty of places you can live if roads are what you want," said Laurie Honkola, a slender, blond woman who runs a fifty-foot salmon seiner. The population is an eclectic mix of old-time fishing families and relative newcomers, many with a high degree of education and intellectual interests. The Orca Bookstore is stocked with volumes that range from botany and chaos theories of physics to poetry and Barry Lopez's cerebral *Arctic Dreams.*

"It takes effort to come live here," said Sheelagh Mullins, a mother of four, part-time vet and former gill netter. Indeed, if you would join the Cordova fishing fleet, you would find it an exclusive club. A permit to gill-net for salmon in the sound, which may only be obtained by buying it from an existing permit holder, currently costs about $160,000; a seining permit goes for upward of $300,000 (or did before the spill); and a new fishing boat may easily cost from $100,000 to $400,000, depending on the kind of fishing. For laying out between a quarter to three-quarters of a million dollars, one buys the privilege of working days and weeks nearly round the clock in dangerous weather, with some prospect of real wealth but usually simply a solid middle-to-upper-middle-class income, with winters off. Given the recent prices of permits, even with substantial state aid in loans, many fishermen are fishing on a highly leveraged basis and, before the spill, were counting heavily on a record-setting year in 1989.

* * *

A striking aspect of fishing here, given the independent nature of those attracted to working the waters of the sound, is the Procrustean degree to which their enterprise is regulated. Technology is deliberately and precisely limited, from the breadth and depth and mesh size of nets to the length of boats and when and where their captains may try their luck. State officials in helicopters count salmon entering the clear-water spawning streams to estimate the optimal harvest.

"We limit efficiency to preserve a lifestyle as well as an industry," said Jim Brady, a biologist with the Alaska Fish and Game Department. "It's a compromise between benefiting more to a lesser degree versus a few to a large degree." In other words, for all their free-spiritedness and macho image, the fishermen have had to embrace regulation and limitation and socialism—the community over the individual—to a far greater extent than the buttoned-down, executive-suited oil industry.

For many of the veteran fishermen who eighteen years ago waged a lonely and frustrating fight against the pipeline coming to Prince William Sound, these are the days of the big I Told You So, but there is no smugness. "We worked to guarantee Coast Guard radar control all the way to the Gulf of Alaska, for double bottoms on all tankers, for strict enforcement of vessel traffic lanes, for pilots on board all the way out [of the sound]," said Ross Mullins, captain of the seiner *Sheelagh M*. "We were too weak, too small, too poor . . . but we were right."

The whole fight is recounted in a fascinating oral history compiled by James T. Payne in 1985, entitled *"Our Way of Life Is Threatened and Nobody Seems to Give a Damn": The Cordova District Fisheries Union and the Trans-Alaska Pipeline*. Ross Mullins came to Cordova in the sixties with a degree in photography. He recalled discussing the oil industry's plans for Alaska in 1971 with his wife, Sheelagh, and "she told me she'd seen what an oil spill was like back in San Francisco . . . [and] I'd better get off my ass and do something about it." He ended up going over to Knute Johnson's basement with an article he had found on the perils of oil in the marine environment. Johnson was mending net and scarcely knew Mullins but said, "Come on in." Johnson's wife, Babe, recalled in a taped interview that it was all very quiet, then there was a sudden explosion of

Knute's voice. Knute Johnson and Ross Mullins stomped upstairs and out the door, and "from then on, all hell broke loose."

The fishermen sued to stop the pipeline from coming to Valdez. To pay lawyers, they agreed to assess themselves a penny for each fish landed. They were opposed by virtually every newspaper in the state, including the *Cordova Times*. At one point in February 1971, according to accounts given Payne, the governor of Alaska, Bill Egan, threatened to embargo all fuel intended for the Cordova fishing fleet: "You guys don't want oil, well, I'll build a goddamned wall around Cordova. You guys don't want oil, I'll make sure you don't get any oil [the governor declined to be interviewed for the history]."

A high-powered oil-industry ad campaign put the village in the position of seeming to oppose national security, as well as a steady and reliable energy supply for Americans everywhere. Even as they pursued this tack, the companies were preparing to sell substantial amounts of the pipeline oil to Japan. Congress later prohibited this, but many newspaper articles during the early 1970s speculated that this was a powerful reason for the companies' picking a marine exit for the pipe instead of an overland route through Canada to the Midwest. The overland route, never seriously considered by the Department of the Interior, was favored in studies by Resources for the Future and the Arthur D. Little Company on both economic and environmental grounds.

Mullins remembered: "We were so discouraged. One night we were sitting in a hotel room listening on radio to a congressional hearing on the pipeline, and this professor from Alaska Methodist University spoke so eloquently against it . . . we didn't even put coats on—it was snowing—and we ran like hell down the street to the hearing room to shake his hand and thank him." That year, 1971, was designated by President Nixon as Fisheries Centennial Year, and all citizens were urged "to support and encourage . . . [the] protection and enhancement of the nation's fisheries."

Reluctantly, Mullins and a few others from Cordova borrowed a car in his native New England and made a whirlwind tour of the East Coast to lobby the media, scientists and Congress to gain support for their case. Generally, they were received politely, but no more. The oral history recalls a *New York Times* editor who sat through a long session, complete with charts of the sound and its fisheries and dismissed them: "When we have a big oil spill in Prince William Sound," he said, "we'll have a handle on the story." And, indeed, in 1989 the *Times* acquitted itself well with its analysis of how the spill happened.

Then there was the interview the fishermen had with Ted Stevens, then and now, an Alaskan senator in Washington, D.C., then and now an unswerving supporter of the oil industry (although he has swerved a tad in recent months, backing a moratorium on offshore drilling in icy Bristol Bay, in southwestern Alaska). "Before we have a chance to say a word," said Knute Johnson in Payne's history, "he says, 'Wernher Von Braun,' you know, the space man, 'assured me that all the technology of the space program will be put into the doggone tankers, and there will not be one drop of oil in Prince William Sound.'. . . And it was so ridiculous that Ross, as much as he loved to talk, it just shut him up . . . and we just got up and left because there was nothing left to say."

In February 1973 the fishermen won. A federal appeals court ruled that the pipeline could not proceed because of a technicality involving the right-of-way across Alaska. The court did not even rule on the much broader environmental-impact challenge to the project, since the techni-cal decision rendered that moot. Quickly, the industry and the Nixon administration shifted the battle to Congress, whose members were panicky about oil supplies, even before the Arab oil embargo that year and its resultant gas-station lines. The crucial vote in the Senate resulted in a 49-49 tie, broken in favor of the pipeline to Valdez by Vice President Spiro T. Agnew. President Nixon signed the project's authorization a week before Thanksgiving. By the time of the Senate vote, the issue had become less a debate over environmental protection than a regional-development issue, with West Coast senators pitted against Midwestern and Northeastern senators over which part of the country would get the pipeline. Ross Mullins recalled there was little to the debate that was profound: "It was not about quality of life, or faith in modern technol-ogy, or national security . . . the state and the politicians just saw the money."

The final summary of the Interior Department's Draft Environmental

Impact Statement for the pipeline stated that the pipe to Valdez would, "in accordance with Department of the Interior stipulations . . . reduce foreseeable environmental costs to acceptable levels." But most of the stipulations that related to tanker safety and cleanup capability were never implemented fully, and some not at all. "The oil companies just whittled them down and whittled them down, and the state and the Coast Guard let them do it," said a Cordova fisherman.

Oil Is a Natural Substance

LET ME MAKE a modest confession here. I don't believe that this oil spill, and oil spills in general, damage the environment as badly as most people think they do. Granted, the Exxon spokesman who announced early on that "oil is a natural substance" was colossally insensitive, as though telling a rape victim that sex is a natural act; but the fact is, more than a decade of close scientific attention to oil spills has shown the marine environment is more capable of degrading and assimilating crude oil than was generally assumed. Studies by a wide range of interests, from the oil industry to the National Academy of Sciences, suggest that biological impacts from large spills are seldom as extensive or as long-term as the public perceives. Marked exceptions are the tragic impact on birds and otters that encounter oil.

Perhaps the worst-case spill in history, from the standpoint of damage to the ecosystem, was from the *Amoco Cadiz* in 1978, when six times as much crude oil as from the *Exxon Valdez* went directly ashore on the coast of Brittany. The extensive salt marshes there are generally a more fragile habitat than the rocky shores and gravel beaches hit in Prince William Sound and less easily cleansed by either nature or man. (The NOAA ranks the nation's coastlines by sensitivity to oil spills, with 10— protected coastal marshes—being the most sensitive. A sheer, rocky headland subjected to high wave energy would be an example of a 1 ranking. None of the shores hit here ranked higher than 8, and at least seventy percent of them ranked lower.) At the *Amoco Cadiz* spill site, which was much harder hit, fishing was normal again in less than three years, mollusks and clams on the bottom recovered in six years. The longest-term impacts were associated with the marsh areas that had been immediately cleaned of oil, sometimes with heavy machinery. Marshes left alone recovered virtually 100 percent within five years, while cleansed marshes took years longer.

The waters of Prince William Sound are mostly hundreds of feet deep, often nearly to the shore, and relatively devoid of suspended sediment, the latter due to the lack of soil washing off its forested, rocky watershed. "That's excellent news for the environment there and bodes well for the fisheries," said James R. Payne, who has done extensive studies for the NOAA on the fate of North Slope crude oil in cold-water environments like the sound. He explained that it seemed almost certain the area would escape one of the biggest long-term problems of oil spills—the burial of toxic portions of the oil in bottom sediments. These can subsequently leach out over time or possibly work their way up the food chain as they are consumed by lower life forms on the bottom, which are eaten by higher predators, and so on. "The two major ways you get that is with sandy beaches, which we didn't have here, and with high loads of suspended sediment in the water [that can take the hydrocarbons to the bottom]," said Payne. He said sediment loads need to be at least 100 parts sediment suspended in every 1 million parts of water for that to happen. Prince William Sound had levels last spring of about 0.4 to 4 parts per million.

As far as toxicity in the water itself, a good deal of the nastiest portions, the so-called light ends of the spill, evaporated in a matter of hours or a couple of days—so much so that the spill probably lost fifteen to twenty percent of its total mass in that period, according to NOAA scientists. The highest level of toxicity scientists at the spill were able to measure in the water was 0.24 parts hydrocarbons to 1 million parts of sound water—that in readings taken in heavily oiled shoreline areas about two weeks after the spill. Readings from a week later found these toxicities reduced to about 0.04 parts per million. "The lowest levels at which we get any effects on sensitive organisms like larval fish and shellfish are from 0.1 ppm to 1 ppm, so I would not expect to see any widespread impacts [in the water]," said John Robinson of the NOAA.

"I think effects from [oil in the water] are going to be very, very difficult to measure. They will be so low, and given the naturally occurring variations in any species, it'll be extremely difficult statistically to ever show the long-term effect on fisheries," said Payne.

No one, of course, argues that the spill was benign or says that Exxon should leave off its beach-cleaning efforts—although the smart money is mostly on nature to finish the bulk of the job. And it seems likely that there are areas where oil has become trapped so deeply within the gravels and boulders of beaches that occasional sheens will be oozing out for years, maybe a decade or more; but the problem will be a lot closer to a nuisance than the catastrophic. "With a few exceptions, I'd be surprised two or three years from now if you can see much impact either analytically or visually," said David Kennedy, Robinson's colleague at the NOAA.

On a national scale, look at what Congress *did not* say about oil and the environment in its December 1988 oversight report *Coastal Waters in Jeopardy*, the first time that body had focused attention on the "pervasive . . . damage and loss" of environmental quality in the bays and inlets and harbors of America's shores from Maine to Alaska. The report devoted one word to oil and none to oil spills or, for that matter, other disasters. Rather, it detailed the degradation of our waters from the constant, everyday flows of sewage, farm fertilizers, sediment from development and deforestation and plowing; toxics deposited by automobiles into the air and onto streets; power-plant emissions, acid rain, herbicides, pesticides from urban lawns, marsh filling for marinas and seaside condos.

Ultimately, it is not by disasters, terrible and mediagenic though they may be, that we lose our natural heritage but by humdrum incrementalism. The Three Mile Island nuclear fiasco, on the Susquehanna River in Pennsylvania, just upstream from my native Chesapeake Bay, galvanized national attention for weeks in 1979 with the specter of radioactive water spreading downstream, poisoning the rich aquatic life of one of the world's most productive estuaries. Years later the Chesapeake is indeed increasingly polluted from upstream in Pennsylvania, but it is no nuclear meltdown that did it. It is decades of inattention to controlling the millions of pounds of cow manure and other fertilizers flowing quietly, incessantly off poorly managed farms, clotting the waters with algal growth, overwhelming the natural balance of life in the great bay.

So what does it all mean? Have we overreacted? If the studies, which assuredly will be done, find no measurable diminution of algal productivity, no untoward elevation of sediment hydrocarbons, nor any bankrupt crabbers, salmon netters and tour-boat operators a year or two hence, then how do we regard the spill of the *Exxon Valdez*? How far is society justified in going to assure it never occurs again?

John Fowles, in his 1983 essay "The Green Man," proposed that while we worry with some justification about our potential for harming the environment, we exaggerate the degree to which nature has already been overwhelmed: "It is far less nature itself that is yet in true danger than our attitude to it." His point, well taken, was that we are resigned to a continuing indifference, even hostility, toward nature unless we comprehend it must be, in part, forever unquantifiable, beyond lucid and rational discussion, unconnectable with any human purpose. "There is something in the nature of nature, in its presentness, its seeming transcience, its creative ferment and hidden potential, that corresponds very closely with the wild, or green man in our psyches; and disappears as soon as it is relegated to a . . . merely classifiable *thing*," concluded Fowles. I would add a corollary—that if we do not treat as a serious crime the disruption of that green man within us, indefinable, unquantifiable though it be, then we will never assign full and proper weight to the damage from events such as occurred on Good Friday in the state of Alaska.

I had been thinking for several days how to express adequately what a lot of the fishing community seemed to feel over the soiling of their sound; how it went so much deeper than even their worst (and probably overestimated) fears of damage to the fisheries.

"There's a deep emotional attachment to our land and water here that's been broken, been violated," said Rick Steiner, a university fisheries adviser resident in Cordova. Indeed, though they are here because of the great numbers of fish, there is far more than the catch statistics operating on the psyches of fishermen these days. Listen to David Grimes and others talk about the beauty in their spring harvest of herring roe, which is deposited like nacreous pearls on the leaves of underwater kelp; about how you can't separate the catching from the spring sights of diving eagles, lengthening daylight, receding snow on the mountain peaks and the phosphorescence of nighttimes: "The herring moving through the plankton blooms like the northern lights, with sea lions and orcas feeding on them, moving like green rockets," said Grimes.

"You gotta crab to know, but it's exciting being out there," said Skip Mallory, a top captain out of Cordova who wrestles giant traps in the stormiest winter weather on the sound for dungeness, tanner and king crab. "You don't just set a pot for crab, you gotta chase 'em, psych out where they're at, and when you get right on 'em . . . well, it's a challenge, and you're doing it in some of the most beautiful waters in the world."

"Yeah, every fisherman has a little area of the sound that's their special place . . . where they've learned the tides, and the holes, the 'lay' of the place," said Laurie Honkola. "You can't homestead the land around here much anymore, but fishermen kind of homestead favorite spots on the water. Over on the Copper River flats there's certain people you associate with certain spots—the Kikerhenik boys, the Softuk boys, the Grass Island boys."

"I feel the social-psychological impacts will far outweigh the ecological ones," said Jim Brady. "There's such an anticipation to fishing each spring, and this [spill] came when hopes were at a peak. These people are businessmen, but they've also gone a long way out of their way to make this something they can deeply enjoy. . . . They identify so strongly with certain beaches, bays and inlets."

"For this to happen here, where there were beaches you'd walk on and who knew if another human being had ever been there before you—just look out there," said Sheelagh Mullins, pointing out her living-room window. "There's a whole range of mountains and no one's in them. They've never been defiled . . . that's why people are here. When it [pollution] can happen here, it makes you feel the whole damn planet must be out of control."

Mullins and many other men and women in Valdez and Cordova would tell me days would go by and they would feel all right, then they would be in the middle of shopping, or doing the dishes, or eating, and they would just cry, they weren't even sure why. I kept coming back to rape as an analogy for what had happened. The physical damage would heal, maybe quickly, but the emotional and psychological trauma, I was convinced, might never go away entirely. I could see a lot of "blaming the victim" occurring: "Well, we all use too much oil, so we shouldn't complain"; "We're all guilty for being part of a consumer society"; and so on. And, thinking about Exxon's controversial efforts so far to clean up the spill—would you assign the rapist to nurse the victim back to health?

One April Sunday, a month to the day after the big spill, a gale was building out over Orca Inlet as townspeople walked, bent against the rain, into the "Home of the Wolverines," at Cordova High, for the first Prince William Sound Day. The signs I had originally seen in Valdez advertised that today was to be Prince William Sound *Memorial* Day, and I had heard some talk about turning it into some good old-fashioned Exxon bashing—"sign of the double cross" and all that. But inside the school the townspeople had turned it into something really beautiful, no memorial, or bitterness. The theme was "Sound Love," and the stage was open "to anyone and everyone who wants to share from their heart to create a vision of and for the sound."

Letters and drawings from kids festooned the walls of the gym—from Auke Bay and Wasilla, from Willow, Juneau, Nome and even San Diego—lots of pictures of whales and otters with big red hearts drawn around them: "Did all the animos diey?"; "Wee love our water to bee clean." I've read so many books about the complexities of man-nature relationships. These kids just cut through all the bullshit—Sound Love. Later we'll teach 'em how much more complicated it all is. Brownie Troop 255 handed out bright bouquets of paper flowers. We had poets, recitations, singing, good and bad; a woman read the Declaration of Independence, no one was sure just why, but it seemed fine.

John McCutcheon, a fine folk singer and songwriter who came all the way from Virginia to help out his friends in Cordova, sang a special composition about the salmon called "Silver Run": "One hundred miles, maybe more, along that living, leaping shore/Oh, we'll cast our nets and dream of better times/All along Prince William Sound, where the silvers run and bound/And our lives meet in the tangle of the lines."

At 2:00 P.M. there was five minutes of silence, which was to be observed by people all over Alaska. Then a pretty woman with the richest long black hair talked to us about the years she and her husband, Ray, spent aboard his old gill netter the *Little Queen*.

"Those were lean years, monetarily," she said, "and the only vacations we had were when we loosed our lines from the dock and left all cares and worries in town and went out for weeks at a time for salmon seining and herring seining. In the summer we would go ashore on the closures [of the fishery] to pick arm-load bouquets of wildflowers to fill the cabin all week long . . . exploring old copper and gold mines and herring canneries . . . hiking up to go trout fishing in beautiful, silent lakes . . . beach-combing and clam-digging for that quick, short burst of pure flavor. And in winters, deer hunts I can never forget . . . winter skies etched in my mind as the most delicate turquoises and palest peaches . . . the pinks and powder blues . . . the flaming oranges and deepening nights . . . the storms . . . when only sure boatmanship forged from a lifetime's experience on the sound, coupled with divine protection, saw us through back home safely."

And she read this poem she wrote to Prince William Sound:

There is a shadow,
dark as death,
lying over this land and sea.
A place where gods are born
and men to privilege see.
I long to enclose you in my arms,
protect you from the large society,
which grunts with hunger for the oil,
and counts the risk of you
too late spent
and then it is too late
and I sorrow to see
your great and final purity.
This is your last hour,

how I would change it if I could,
perhaps instead your waters should.
But nothing can remake your essence
once they begin to take
their greed and poison:
Your death will be their fate.
Stripped of your trees,
your beaches black and lifeless,
I will remember the day
it was not so.
And I will love you
all the more.

It was fifteen years ago she wrote that, Christine Honkola said the next morning as seven children, some hers, played around her cheery mountainside log home. It was the pipeline that brought her here, she says, "because when it began, I didn't feel like Anchorage was a place any longer for a woman alone. . . . It seemed like a rape a day there." Her husband, out on the spill cleanup, has heard fourth hand that an Exxon skimmer filled up, had no place to unload and was told to dump it back and skim it up again—apocryphal perhaps, but it will go into the lore that's building. The poem she read yesterday seemed remarkably prescient, I said. "Well, we all felt it was inevitable, but it hasn't made what happened easier to take," she said, adding, "If any message goes out to the world, it's that people have to have more say for their concerns about environmental protection, because they are the ones who have the most to lose, not the corporations."

What was lost? Much that can never be adequately articulated, if Fowles is right. Ultimately, it is the poets and singers, and not the scientists, who will come closest to telling us. The true verdict will be rendered in art, in the gut, not in the lab, and I suspect it will be less exonerating than the ten-year follow-up study that is reported at the 1999 Oil Spill Conference in Houston or New Orleans. And to the extent we do not sense that, heed it, we will be vulnerable to future oil spills. We need the numbers, the biological and financial and legal accountings of what happened here and who was responsible, but those alone will not save us from another Joe Hazelwood. A bigger voice in the future of Alaska for those with Sound Love might, though, along with a fuller understanding of what the spill meant to them.

The Future

EXXON BY SUMMER'S start has 9,000 cleanup workers, many camped in troop ships anchored around the sound. All the beach cleaning is finished. By mutual agreement the state and the oil company now refer to beaches as having only been "treated," which is a tip-off to how the cleaning, whoops, treatment is going. Oil is lodged so deeply in the rocks and gravel that sometimes hours after the surface is, ah, treated, a high tide refloats the deep-down brown goop and reoils the area. Exxon claims, by late June, to have treated forty miles of beach within the sound, about a tenth of what was hit, but some in the state say those figures are suspect. New cleanup methods continue to be tested—chemical dispersants, oil-eating microbes—and the phones continue to ring with helpful hints, such as using the cheese packets from Kraft macaroni and cheese dinners, also duck feathers and peat moss. One wag suggests using sea gulls, which, he notes, are plentiful, highly absorbent and can be wrung out and reused two or three times. It promises to be a long summer.

Now the debate is shifting, both in Alaska and around the country, as to what the impact of the spill will be on environmental-protection and energy policies.

On an Alaskan airliner, I come across an ad in the flight magazine that no doubt was printed before March 24 [1989]. The ad is for Unitech, a supplier of oil-spill-cleanup equipment and supplies, and it features a fetching model in an attractive dress seated amid a small cluster of dispersants, sorbents, protective gloves and boots: "Unitech . . . because accidents do occasionally happen." In cleanup capability, the stuff in the picture is about one step up from Bounty towels. Still, on March 23 it would not have seemed remarkable; now it is absurd, laughable—outrageous and insulting.

Clearly, the frame of reference has shifted for us all regarding how bad an oil spill can be. And though what happened here is often described as worse than anything ever imagined, it was nowhere near the worst case. For example, Hazelwood tried for more than an hour and a half to power his dangerously unstable and badly rent ship off the reef. Had he succeeded, and had it broken apart or sunk, fresh oil might have streamed to the surface for months. The weather was extraordinarily cooperative in aiding cleanup. In December the days last only five hours, and storms and snow often cut visibility almost to zero for days at a time. Or what if it had happened a month later, just as the $50 million salmon run was homing in on the sound? Or what if the winds that normally prevail had been blowing from another direction, toward the marshes of the Copper River, used by an estimated 10 million birds annually as resting and feeding areas?

Additionally, the spill has dispelled any doubts that our capacity to transport oil in large volumes has far exceeded our capacity to clean it up, both technologically and managerially. It is instructive one day to tour the high-tech factories of oil extraction on the rich fields of the North Slope; the next day marvel at the sophisticated instrumentation on the bridge of a supertanker; then a day later slip and slide across a beach as grimy, discouraged men and women with paper towels try to wipe oil off rocks while back in Valdez bureaucrats and corporate officials and their lawyers argue over what to do. Even at Sullom Voe, the oil terminal in the Shetland Islands considered to have the world's best spill-response capability, a recent practice spill simulating a loss about an eighth the size of the Valdez spill resulted in some (theoretical) oiling of beaches.

Could a catastrophic spill happen as easily elsewhere along the nation's coasts? Arguably, it was less likely to happen at Valdez than most places, many oil-spill observers say. "Because of the immense pressure to take whatever safeguards necessary to get on with the pipeline, more was done to anticipate the problem in Prince William Sound than is likely to ever be done anywhere else," said Saunders Hillyer, a lawyer who fought the pipeline and now works for the Chesapeake Bay Foundation, a group based in Maryland that watchdogs oil pollution and a wide range of other water-quality issues.

That sentiment was seconded by a congressional staff member who works closely with oil legislation. The spill-response plan at Valdez, for all its failings, is still much more than most places have, said the staff member, adding, "Most other places, if they have any contingency plan, do not include response planning for tankers after they depart [the dock or harbor]. Technically, every place else on tanker and barge routes is covered by the National Response Plan, but that involves basically a list of what agencies respond . . . As far as detailed regional planning and prestaging of cleanup equipment, there is not much."

And a retired Exxon employee who now reviews contingency plans around the country as a consultant said that "most are not worth the paper they're written on . . . They are lists of phone numbers, not action plans." Similarly, he and others said, Valdez was relatively progressive in having a plan that divided the sound into areas of greater and lesser sensitivity to aerial spraying of dispersants. But even that plan dissolved in controversy when it came to the real spill.

"The real issue is that the contingency plan left the state in a reactive position," said Dennis Kelso, the Alaska environmental conservation commissioner. "We have to reevaluate. . . . Is it realistic to have a party like Alyeska, which has consistently been opposed to additional environmental-protective measures, propose and implement such a plan?"

For most coastal waters around the nation, the Coast Guard is the first line of defense in preventing and responding to oil spills, with authority to require and review contingency plans, set and enforce vessel safety standards, coordinate cleanup efforts and send national strike-force teams to spills. That agency got a schizophrenic combination of low marks in preparedness and sympathy from most experts. "Valdez probably started out being a model terminal, but the Coast Guard never took control," said Pete Johnson of the Oceans and Environment Program of the congressional Office of Technology Assessment (OTA). "They left too many decisions about running the place up to industry—oversight was the missing link."

The Coast Guard's history in recent years has been one of level budgets and new responsibilities, such as drug interdiction. "They have not done a good job anywhere with oil transport," said Johnson. "But I think it's a question of having to set priorities. They feel they are primarily charged with saving life at sea. If you check, you'll find they do a better job with safety on human-life-threatening cargoes like LNG [liquid natural gas] tankers." The Coast Guard has closed eleven of its marine-safety offices and merged its only East Coast oil-spill-response team, in North Carolina, with one in Mobile, Alabama. It never followed through on plans earlier in the decade to pre-position response equipment at eight sites around the coasts.

During this period a number of special funds, totaling hundreds of millions of dollars, have been created through taxes on offshore drilling and pipelines to handle various impacts of oil development—but none of this has ever been earmarked for spill response. Legislation to remedy this is pending in Congress, and there are hopes the Alaska spill will lend it impetus. Budgets for research devoted to oil-spill-cleanup technology in the Coast Guard, the EPA and the NOAA, which averaged less than a million dollars per year until recently, have been cut even further, according to the OTA, and a government spill-technology test center in New Jersey has been closed.

"In general, there is no adequate response capability to a spill of [the *Exxon Valdez's*] size," said Johnson.

Just that point has been driven home in recent weeks by Governor Cowper and the Alaska state legislature, which are urging Congress to put a moratorium on proposed oil-development leasing in Bristol Bay. It and other potential drilling areas here, like the Chuckchi Sea, are even more remote and more hostile environments than Prince William Sound, opponents of the drilling have noted. The fishery in Bristol Bay is much larger than that in the sound, worth an estimated $1 billion a year.

Alaska's congressman, Don Young, and the state's two senators have backed off in their support of continued leasing of offshore areas for oil exploration. Even oil-industry officials have said privately that they think further offshore exploration in Alaska's icy northern waters may be shut down, with a possible federal buy-back of existing leases.

The state legislature recently repealed a handsome tax break the industry had enjoyed for years on certain oil fields, and the state has also reacted to the spill with an emergency order implementing tough, new safety measures at Valdez on both transportation and spill response. At Valdez, ships are now accompanied on their passage through Prince William Sound by new emergency-response vessels, each one carrying 4,600 feet of boom and with the capacity to hold 4,000 barrels of oil. Skimming vessels and barges are on continuous standby in the sound, midway on the tankers' route. Pilots now guide vessels past Bligh Reef, although not to open water, as they once did. Captains and crews undergo alcohol testing before departure. The state disputes whether Alyeska has enough cleanup and prevention equipment in place.

But the big test looming for the industry [in 1989] is the decision over the fate of something known as Anwar. This is a 100-mile stretch of Arctic coastal plain, 1.5 million acres, within the 19-million-acre Arctic National Wildlife Refuge (ANWR). The industry thinks it may contain another oil field as rich as Prudhoe Bay, with its billions of barrels, or at least a find in the hundreds of millions of barrels. And it is close enough to current North Slope oil fields to be conveniently hooked to the existing pipeline down to Valdez. According to environmentalists, the portion of ANWR where Congress has said it would consider oil leases is the heart of a magnificent wildlife area. To make it into another North Slope oil field would be to violate in fact and spirit the entire concept of setting aside such a vast area as unspoiled wilderness, they have said.

Until the spill, ANWR was on the fast track in Congress for leasing and exploration, even the bitterest foes of exploration concede. Now even its most ardent supporters in the industry and the Bush administration admit that it will be delayed for at least another year, and environmentalists think now that there is a real chance to keep the refuge sacrosanct.

Perhaps the most far-reaching ripple from the *Exxon Valdez* spill, however, will be a redirection of America's energy policy, always a lively topic in a nation where six percent of the earth's population consumes twenty-five percent of its energy. Is it "energy independence" we pursue or the continuation of history's most luxurious and consumptive lifestyle?

The largest untapped reservoir of oil in the United States is not beneath the Alaskan wilderness but in the nation's fleet of gas-hungry trucks and automobiles, says Ralph Cavanagh of the Natural Resources

Defense Fund. The trick, say conservationists, is getting the public to think of fuel-efficient cars as oil wells you don't have to drill, of high-efficiency air conditioners as power plants you don't have to build. "You don't get Exxon's attention by boycotting Exxon—you get it by boycotting gasoline [through conservation]," said Peter Dykstra, a spokesman for Greenpeace, which declined to join the wave of organizations boycotting the oil giant after the spill. Such perceptions were all but impossible for environmental groups to foster during the eight years of the Reagan administration's reliance on decontrol, free markets and stepped-up oil leasing on public lands and in coastal waters.

George Bush, a former Texas oil man, scarcely seems ready to hop on the conservation bandwagon—"I have to harvest the resource—that is, sell oil, coal . . . those kinds of things," his interior secretary, Manuel Lujan, told Alaskans recently—but nonetheless the prospects have never seemed brighter in this decade. "A year ago, I said that with $1-a-gallon gasoline, we were never going to redirect energy policy," said George Miller, Democrat of California, whose congressional subcommittee has been holding hearings on the spill. "[But] we may now be in a situation where people are willing for the government to get ahead of the market . . . I just believe the politics are changing."

Automobiles are by far the largest users of oil, burning nearly forty percent of all the oil consumed in the United States. They now average a mile per gallon less (26.5 mpg) than they were supposed to under a federal conservation law adopted in 1975. The Reagan administration relaxed the standard and called for the law's repeal. How much difference does a mile per gallon make? Increasing the fleet's average mileage by 1.7 mpg would save 3 billion barrels of oil over the next thirty years—the equivalent of pumping a major field beneath ANWR, according to Jan Beyea, a senior energy scientist with the National Audubon Society. And that is only a start. The Office of Technology Assessment calculates a 33 mpg average is possible by 1995, "without performance loss or the need to move to smaller vehicles." Higher consumer costs for the re-designed engines, transmissions and bodies that would permit this higher mileage would be recovered by gasoline savings, according to the OTA.

America's automakers are marshaling arguments to resist the calls for more efficient cars, which they say restrict consumer choice and add to highway deaths by forcing people into smaller cars. They also feel Japanese car makers would have a competitive advantage.

Perhaps more important than the energy costs of burning petroleum are the health and environmental impacts. On the front page of the April 3 [1989] *New York Times* was an ironic juxtaposition of articles. One detailed the damages of spilled oil in Alaska. The other announced that EPA scientists were beginning to wonder whether the nation's growing air-pollution problem was not simply intractable and too expensive to pursue by the traditional method of adding more "widgets on smoke-stacks."

Vehicles on the road are the largest source of the emissions that result in smog. Even though carmakers have controlled tail-pipe emissions sharply since 1970, the number of vehicles on the road has increased by seventy percent—and people are driving them farther. Accordingly, much of the nation still fails to meet the standards set by the landmark Clean Air Act of twenty years ago. President Bush's new clean-air proposal, announced in June, does not go nearly far enough to correct this, many environmentalists feel.

In addition to the serious health effects of smog, burning more petroleum and other fossil fuels (coal and natural gas) contributes to the greenhouse effect. Combustion of fuels is gradually increasing the amount of carbon dioxide to levels never before present in the earth's atmosphere, and this in turn is accelerating a global warming trend. The results, during the next several decades, will range from unprecedented rises in sea levels, threatening coastal cities, to major adverse shifts in grain-growing areas.

"We talk a lot about the impact of the spill on Prince William Sound, on fishing, on Exxon's finances . . . but I think the biggest impact's been psychological," said Dave Cline of the Audubon Society as we talked in his Anchorage office. "I can't explain it, but I don't know of any other environmental tragedy that's had this impact on people . . . such feelings of rage, sobbing. Maybe it's just this spill on top of syringes' washing up on beaches, the earth's ozone layer unraveling, animals in the oil . . . just a feeling that things have gotten out of control. Maybe ANWR is where we are going to make a basic choice about which direction this country goes."

Globally and nationally, there are compelling economic and environmental arguments for curtailing our use of oil. But what of Alaska, which derives close to nine-tenths of its public revenue from the pipeline and the industrial elixir flowing from it? Attempts to diversify Alaska's economy have largely been overwhelmed by the prospects of drilling in ANWR, squeezing extra years out of the existing Prudhoe Bay field and pumping to Valdez the huge quantities of natural gas that also exist on the North Slope (but are currently uneconomical to recover because of the low level of world prices). "Long-range economic planning up here," said Gregg Erickson of the governor's office, "usually means figuring how to get through the winter."

The hard truth is, economists like Erickson and environmentalists like Cline agree, that there is not now, nor ever will be again, another revenue machine like the Trans-Alaska Pipeline. "Alaska is lean country, a vast amount of it is not that productive, not that habitable," said Cline. "Our bias is that one of the greatest things in Alaska's future is its wild scenic value. Oil has created unrealistic expectations of the future. In other words, the future of Alaska may be less, not more—which does not mean there aren't tremendous opportunities in tourism, fisheries and other areas, but nothing will equate to oil money."

The giant Prudhoe Bay field is already entering its inevitable decline, it was clear on a visit to the North Slope. The industry has recently invested nearly a billion dollars in new recovery facilities to squeeze more crude out of the underground caverns by injecting saltwater; also on reinjecting the natural gas that comes to the surface with the oil. Despite all this, within a decade it seems likely that the flows through the pipe will have declined to as little as half a million barrels a day, versus the current 2 million. Even if the go-ahead to explore and drill ANWR came today, and even if oil were found there in commercial quantities, it would probably be twelve to thirteen years before it was flowing to Valdez. Thus Alaska is virtually guaranteed to have a sharp decline in revenues in the foreseeable future. "The question then," said Gregg

Erickson, "is, will people accept not being the richest state, or will they leave?"

One answer to that is Loretta Lewis. She is one of that hard and substantial corps of Alaskans who were here before oil and will likely be here after it. In a state almost totally dependent on oil, they are less dependent on it than most lower 48-ers. I met Lewis waitressing at the Pipeline Club, which serves as mailbox, friendly refuge and watering hole to merchant seamen. Until recently, Joe Hazelwood was just another good customer here. Lewis had come down from Delta Junction, "home of the wind, honey, 'cause it blows like hell and blows silt off the delta all year long."

"I support a dog team, and I needed income," says Lewis, pointing to her T-shirt: "Yukonquest 1,000 mile sled dog race." Lewis says she has been sticking with the same bloodline for six years now, and it brought Jeff King, who bought her dog Molar for his team leader, a first place in the 1988 Yukonquest. She is hoping to latch on to some of that good Exxon money, living, meantime, in her battered Dodge pickup with handicapped plates ("bum knee, quack doctor," she says). Her boyfriend lives there, too. "He's a lot younger than me—I'm fifty, and he's a miner," she says. "He gets discouraged, but I tell him, 'Look, you came here to mine, then MINE—win, lose or draw, you're doing something you like to do.' "

There's no water and no electricity in Lewis's sixteen-by-sixteen Quonset in Delta Junction, "and I will say the last year's been hard on everybody, with the economy down, but I don't require much." Of the oil pipeline, she says, "We have to have something for our young people. . . . I've seen friends bust their tails, practically eat out of garbage dumps, trying to make it logging and trapping, and they couldn't. I'm no environmentalist, and I don't want to pave the earth, either—I just think there must be some way we can live on this earth and not hurt the animals anymore. My family, we go out in the woods, we leave it like we found it, simple as that."

Lewis came here from California, where she was a letter carrier for the Post Office, putting her kids through school before heading north. For a while she ran her half brother's hotel in San Pedro, where seamen laid over. "And Exxon says they didn't know they drank?" she says. "I coulda told them that. I used to think, 'My God, do they do this on the ships?' "

What Lewis likes about Alaska, she says, "is the pureness of the place. . . . I had been changing my mind about ANWR even before this spill, because I just don't think the companies'll take care of things." She's trying to diversify her own economic circumstances, she says, learning to paint on birch fungus and doing woodcrafts out of diamond willow and spruce burls. She's got an old Indian chief teaching her how to make model birch-bark canoes. "It's tough up here, but where else can you go that the grizzly bears come down to the stream and buffalo roam through your yard?" she says. "I'm a quarter Navajo, and those buffalo just give me a thrill. I tell ya, it's like that show, remember, *The Naked City* . . . 8 million stories—someday I'm gonna write about it when I get to where I can't do things—oh, now don't take that last bit personal." She taps my notebook.

Alaska's a good state, says Lewis, limping a bit as she heads for new customers—"it's a crapshoot, hot or not, boom or bust. I wouldn't go back to the lower 48 for nothing."

Winter comes early in Alaska. By late September the Exxon army that swelled to 12,000 troops has left the windswept sound to the bears and eagles and a modest, state-funded effort by local fishermen to pick up oil-soaked seaweed, which still litters the intertidal zones of the coast. Of 1,090 miles of sullied shore, 1,087 miles have been "treated" or "stabilized [new gravel dumped over the old]," according to Exxon. The state of Alaska says 1,000 miles of this needs more cleaning. Costs to Exxon so far have exceeded a billion dollars, with about a third of that being covered by the company's insurers. Lawsuits filed against Exxon at last count came to 145.

Scientists will also be prowling the sound's beaches this winter, documenting the oil's fate as it degrades and hardens, by spring, to the consistency of a highway. Already they are raising a troubling question: To what extent was the mammoth cleanup effort misguided? It seemed at the time that we had to do *something*, didn't it? There was also a satisfaction—a need for it perhaps—in witnessing Exxon do public penance, and the months-long effort did provide a vehicle for disbursing hundreds of millions of dollars through the Alaskan economy, workfare being more acceptable than welfare or hush money.

But there are a few small studies back from the beaches that indicate recovery times for the environment in some spots will be longer than initially predicted—not because of the oil but because of the physical impact of being scrubbed and hosed and trampled. "I'm really striving very hard to improve people's understanding of [the benefits of] doing little or nothing in such situations," says David Kennedy, the NOAA spill veteran. "There is often a strong case, environmentally and economically, for leaving it alone." Even with the animals? I ask, knowing the respect he has for Jessica Porter, the deeply caring vet at the bird-recovery center and a neighbor of his back home in Washington State. "Well, in Canada, it is federal policy not to assist in any way with bird recovery in spills, given the low survival rates and the high costs," he says. Kennedy expects the wrangling over cleanup to begin anew with the spring. There's been talk about making Exxon clean the sound to prespill levels, "but we just don't have the technology to do that," says Kennedy.

Since Valdez, a number of states and localities have conducted spill drills of their own, all theoretical, of course, and generally with lots of advance preparation. I haven't heard of anywhere the parties involved came out and said the next day, "If the big one happens here, it's good-bye to the [insert your favorite marsh or river here]." One wonders; would it push us to improve upon prevention, or to rethink the limits of where we drill for oil, if authorities had to admit that we are, in big spills, most likely to be working without a safety net?

There are, to be sure, significant changes coming in oil-spill protection as a result of the *Exxon Valdez* catastrophe. "It has put oil-spill reform among the top environmental priorities of Congress," says Will Stelle, majority counsel to the House Subcommittee on Fisheries, Wildlife and the Environment and a veteran of several years on oil legislation. The traditional focus on liability and compensation of victims has been broadened markedly, he says, to force better prevention and cleanup response, in a major spill bill now before the House. In the bill, Congress is finally addressing the issues of contingency planning, both aboard vessels and at marine terminals; also scrutinizing whether manning requirements and vessel design should be changed to make spills less likely. Finally, the bill as drafted would beef up the power and capability of the federal government to clean up spills.

The oil industry has voluntarily created its own new response organization, with initial funding of at least $70 million. This will create five regional centers, staffed around the clock, each equipped to respond to a spill of 200,000 barrels, according to industry spokesmen. David Kennedy describes this effort as "pretty significant—it is also going to put several million into research." But, he says, "I don't see any radical changes, and I think that's what will be needed to markedly improve our ability to clean up big spills."

In Cordova, many of those who profited mightily from Exxon cleanup money are now known as spillionaires, a term that may be said jokingly or bitterly, depending on whom you are talking to. Lots of the spillionaires have taken off for Hawaii and other warmer climes until fishing resumes next spring says Rick Steiner, the University of Alaska Sea Grant agent in Cordova. He and fisherman David Grimes have been making forays to other coastal states all over the country, telling anyone who will listen to learn from Prince William Sound's experience, not to be complacent. Once, at a press conference in North Carolina with a number of local dignitaries, guards at the marina thought the two guests of honor were too scruffy looking to admit. Grimes and Steiner jumped a fence across the harbor from the press conference and swam over. Hauling themselves out, dripping, they apologized to the assembled officials for being late, but it was a long swim from Alaska.

I watched an oil barge recently, pushed in front of a tug through the golden autumn salt marsh, move up a small tidal river close to where I live near the Chesapeake Bay. Where U.S. Route 50 crosses the Nanticoke River, the barges, nearly as long as the river is wide, must pass through a drawbridge, nearly brushing its massive concrete supports as they do. The tides are tricky at that stretch, and getting through is a much more difficult maneuver than anything the supertanker captains in

Alaska routinely do on Prince William Sound. The bridge tender says that this past summer one of the tug captains miscalculated the current and smashed the bridge with a full load; thank God, the dolphins (clusters of huge wooden poles sunk in the river bottom) fended him off enough so that he didn't crack her open.

If ever there was a place you don't want an oil spill, it is there. Each spring and summer, so many striped bass are drawn to that very stretch of river to spawn along its swampy fringes that their success or failure measurably influences stocks of the fish throughout the Chesapeake Bay and the length of the East Coast. Shad, river herring and white perch also use the same region for reproduction, and downriver is a productive oyster- and crab-harvesting industry. It is also one of the most unspoiled and aesthetic waterways left on the mid-Atlantic seaboard. An ex-crewman told me that about six years ago, the barge company began using bigger barges on the upriver run—so big that the captain's visibility going through the bridge, with all those deck pumps out there 300 feet in front of him, is almost zero. It's a disaster waiting to happen, he said. All it would take to make it safer is inexpensive modifications to the tugs' wheelhouses to afford better visibility—or a return to the smaller barges. But there haven't been any spills, and that, he said, seems good enough for the company.

So I will watch eagerly the passage of the *Exxon Valdez*–inspired spill legislation through Congress, and dutifully applaud the new response centers going up around the country; but I will also watch those barges going up and down my favorite river, and watch especially hard for signs that we are becoming more caring toward bass and marshes, otters and salmon and the like.

PART V

Power of the People

A common saying among the members of environmental nonprofit groups is that their goal is to put themselves out of business. If they rid the world of environmental dangers, then they can slip back into the mainstream, becoming "normal" citizens again.

Yet the image of self-sacrifice that clings to the environmental movement has become tarnished during a decade when environmental organizations have started to mimic the large, professionally run corporations they often battle. Currently the debate rages within the environmental community over the development of many of its members.

To be fair, environmental problems have not ceased to exist, and the closing down of operations is not a likely choice in the future. But an organization's size—and the inevitable bureaucracy that comes with it—its alliance with business interests and the contrast between behind-the-scene lobbying and on-the-scene protest have created much debate within the community. There have been complaints that organizations have become overextended or that creating coalitions with business interests to solve problems will increase the pressure to compromise environmental standards.

The articles that follow investigate two of the most pioneering and influential environmental groups of the eighties: Greenpeace and the Environmental Defense Fund. These and other nonprofit organizations have been responsible for many of the past decade's environmental victories,

and their structure and style will determine the shape of
the movement in the years to come.

In the 1990s nonprofit groups are finding that they have
more power than ever before. A coalition of such groups
formed in 1990 and 1991 to fight a piece of legislation in
Congress that allowed the administration to negotiate a
free trade agreement with Mexico without congressional
input. While the bill passed despite their opposition, the
amount of press and popular response generated by the coalition surprised many people in the government. At an
international meeting in Bergen, Norway, in 1990 non-
governmental organizations were included for the first time
on the negotiating teams of some governments. That created what is now referred to as the Bergen Process, under
which the expertise of nonprofits is drawn upon by governments to strengthen their understanding of environmental
issues.

But just when their influence seemed to be at its height,
environmental organizations were hit by the recession of
the 1990s. Since much of their money comes from direct-
mail efforts and member contributions, many organizations, including Greenpeace, have had to retrench under
the hard times, a sign that even the environmental movement is subject to economic forces.

The structure and form of environmental organizations
will continue to affect the direction of America's environmental movement. Nevertheless, as Tom Horton states,
long gone are the days "when fate and karma still played a
bigger role than direct-mail fund-raising and satellite communications."

11

The Green Giant

Tom Horton

*T*he task at hand this morning is a tour of the Amsterdam command post from which Greenpeace, the largest environmental organization in the history of the world, runs its modern global enterprises; but my guide can't resist a story from the old days, when fate and karma still played a bigger role than direct-mail fund-raising and satellite communications: "We were on our way to the Canadian ice to protest the seal hunt," recalls Steve McAllister, Greenpeace's deputy international director. "It had been a rough sea voyage from the start, storm after storm, and now the ice was getting thicker. The old *Rainbow Warrior* [the group's flagship] only had a truck motor, and it didn't have the power to break through. . . . For a while we tagged after a freighter," McAllister continues. "Then, twenty miles from the hunt, we had to leave the track. We were stuck. No forward, no backward. The journalists on board to record our confrontation with the seal clubbers were getting surly. We'd promised a major action, and now the whole campaign, with the reputation of the organization riding on it, was going to be an embarrassing bust.

"Suddenly, the gray skies parted. A sunbeam came down. The ice cracked into a river of open water leading right toward the sealing grounds, and a rainbow formed over the river. We stood there, dirty and exhausted and happy, and thought: 'Too much, God, you didn't need to throw in the rainbow.' "

That was scarcely a decade ago, but it seems a light-year, McAllister says. Back then about 100,000 people in the United States and Europe belonged to Greenpeace, donating about $1 million annually. Both supporters and money have been growing phenomenally ever since, doubling every year or two. Revenues this year, worldwide, are projected at around $160 million, and the number of contributors is nearing 5 million. The U.S. office, Greenpeace's largest, has about 2.5 million of those, with projected 1991 revenues near $60 million. Worldwide, Greenpeace now has offices in twenty-three nations.

These numbers are made all the more impressive by the fact that Greenpeace, virtually alone among major environmental groups, solicits no money from wealthy high rollers, corporations or foundations. Its financial clout, global spread and huge contributor base make Greenpeace potentially as great a force for saving the planet as any organization that ever existed. It has gotten major credit for saving the great whales and recently played a leading role in saving a whole continent— Antarctica. This summer all nations with a claim to the frozen but resource-rich continent agreed to a fifty-year moratorium on exploitation of oil and minerals there.

It may seem incongruous, therefore, to note that as Greenpeace looks toward the next century, it is in a state of turmoil about its future, agonizing over whether success has dulled its cutting edge. Is it poised to become a truly global force, a sort of activist, ecological United Nations, or is it in grave danger of becoming overextended? Is it becoming too comfortable and bureaucratic to take the risks that have been Greenpeace's stock in trade? Can the results it is getting—muddled in some areas—justify a budget that is projected by century's end to exceed a quarter billion dollars?

To put the situation in perspective, turmoil has existed in almost every year since Greenpeace launched its first voyage on behalf of the environment twenty years ago this September. Crises, raging internal debate and major philosophical and ego clashes are somewhat the nature of the beast. On paper, Greenpeace is organized straightforwardly enough, as a group of member countries, all under the umbrella of an international headquartered in Amsterdam. Probably few of its supporters in the United States or other nations know that a quarter of every dollar they contribute goes directly to the international office. The mechanism that holds this together is the international's control of the trademark, the name Greenpeace. The power to bestow or withhold that name is worth about $40 million annually to the [Amsterdam] office.

In reality, Greenpeace operates more as a loose confederation of semi-autonomous states, organized around what the group calls campaigns. They run the gamut—ocean ecology to nuclear nonproliferation; energy policy and the ozone hole to the preservation of Antarctica; toxic chemicals to Taiwanese overfishing of the high seas. It is the campaigners who climb the factory smokestacks, man the Zodiac inflatable rafts and parachute into nuclear-blast zones. Campaigns are the essence of Greenpeace, and so far they have tended to drive it more than any centralized planning process.

In addition there simply is no guidebook for running Greenpeace. There has never been anything like it. After twenty years of existence, the group's motto (if it could agree on one) might well be a sentiment often expressed by Greenpeace's managers: "We're running a big experiment here." There is, in short, no role model, no blazed trail, no telling what comes next, for a rich, planet-saving, lawbreaking, in-your-face, semi-anarchistic, multinational corporation.

Back in the mid-1980s, when he was skippering a little red ketch, the *Aleyka*, and harassing polluters along the East Coast of the United States for Greenpeace, McAllister used to hold forth about the virtues of never allowing bureaucracy a foothold in the organization. No one should be encouraged, by job benefits or a livable salary, to stay around more than a few years—Greenpeace should be like a hitch in the army; do your patriotic duty for the planet, then move on. Years later, not only McAllister but a large number of Greenpeacers from that era remain with the organization. Salaries have risen substantially, though the money still seldom exceeds $45,000 a year for top officers, and pressure for benefits is increasing now that Greenpeacers are aging, marrying and having babies. For those in command of today's bigger, richer Greenpeace, the possibilities for effecting change can be heady stuff. McAllister notes that his budget for running campaigns in 1983 was $650,000. Now it is in the tens of millions.

There are days, McAllister says, when he wonders how he fits into Greenpeace anymore: "There's no way, out of the thousands of job applications we get, Greenpeace would hire someone like me now." He grew up poor on a Vermont dairy farm, fired artillery in Vietnam and came up through the ranks in Greenpeace—diesel mechanic, body on the protest lines, toxic-chemical-campaign director and eventually coordinator of campaigns for Greenpeace USA. This is the first time he has had to learn to type.

This morning he has tapped a few keys upon entering his office in the beautiful old five-story building Greenpeace International owns along one of Amsterdam's central canals. Immediately, the computer prints out messages from colleagues as far afield as Australia, Washington, Rome and a ship somewhere in the Pacific. Another few keystrokes and McAllister can access environmental news from any of the world's major news wires and many of its minor ones. If he wants to research anything from dolphins to the environmental impact of the Gulf War, he can tap into one of the world's best environmental databases. It is all part of Greenlink, a satellite communications network that includes all of the group's far-flung ships and offices, including a base in Antarctica. It is the kind of system usually accessible only to large governments or the more sophisticated multinational corporations.

"Without the Greenlink we could not exist as we are now, spread around the globe," says Dick Dillman, Greenpeace's resident communications guru in San Francisco. "We could not possibly coordinate the actions we do."

With Greenlink—and Greenbase, its companion database—Greenpeacers all over the world can within minutes educate themselves on issues and on what the organization is doing anywhere. A media specialist at the site of a campaign action can send a press release instantly to hundreds of fax machines around the world—from a portable cellular phone in many cases. Dillman's next project is live video transmissions from ships at sea. "Anybody can do it from dry land, but with ships you have to stabilize the antenna," says Dillman. "We're close." Greenlink is not infallible—it can scan the wire services for key words and down-load any story containing them, but on Monday mornings it has been known to turn up a lot of football stories if the Miami Dolphins played that Sunday.

McAllister shows me a plastic toy he keeps as a memento, a tiny Godzilla that will click and clack and move randomly when you wind it up. It was only eight years ago that Greenpeace's communications equipment was so primitive that he and others wandered the Bering Sea for two weeks trying to find and harass a Japanese fishing fleet. One of the crew, Jim Henry, put the plastic Godzilla down on the charts in desperation, wound him up, and they followed the course thus set. Bingo.

These days a whole side of one floor of the Amsterdam building is allocated to Greenpeace's marine division, which is needed to staff, supply and coordinate the movements of a fleet of eight ships up to 200 feet in length and valued at more than $12 million. They are deployed from the Antarctic to the Baltic Sea, from the Gulf of Mexico to the Great Lakes. Most of them carry satellite communications and facilities for transmitting pictures directly to the news media. One carries twin helicopters aboard.

A 1984 addition to the fleet, the *Beluga*, carries a state-of-the-art mobile lab that the U.S. Environmental Protection Agency would give its eyeteeth to have—it can sample continuously for a variety of water-quality conditions and look for toxic chemicals while moving down a river or through a polluter's discharge plume.

Another recent addition, the 181-foot-long *Rainbow Warrior*, can steam at sixteen knots, a far cry from the original with its old truck engine, chosen because it could fit down the ship's smokestack and save scarce installation money. The original *Rainbow Warrior* was sunk, in 1985 in New Zealand, by French-government agents in order to stop its voyage protesting French nuclear tests in the South Pacific. One Greenpeacer was killed. The resulting international scandal threatened briefly to bring down the French government, which finally agreed to pay Greenpeace more than $8 million in reparations.

Another section of the building holds the organization's "suit-and-tie campaigners," a small diplomatic corps of experts in international law and foreign policy. They represent Greenpeace in treaty conferences and conventions that regulate everything from whaling to ocean dumping. This illuminates a side of Greenpeace that is under-appreciated, especially in the United States. In Europe, Scandinavia and several less developed countries, Greenpeace has a prestige and access to government far greater than any environmental group enjoys in the United States. In international treaty negotiations, the group's influence can extend literally to writing the policy positions of some governments.

PhDs—some full-time employees, some hired as consultants—are cropping up frequently in Greenpeace, whose idea of scientific research used to be distributing fact sheets to the press after stuffing inflatable rafts up a polluter's discharge pipe. The organization recently has produced definitive texts on global warming and on military affairs. It has its own books division in London. It also funds an eco-toxicology lab on the campus of Queen Mary's College, in Britain. Greenpeace has also started its own environmental education curriculum in seventeen schools in North America, Europe and the U.S.S.R., an outgrowth of an idea McAllister had several years ago to create "Green Teams" as a sort of ecological alternative to the Boy Scouts and Girl Scouts. With a link to Greenpeace's database by a computer in Boston, schoolchildren are shown how to identify local environmental issues and formulate plans to solve them.

A legal department is vigilant in protecting the name from trademark infringement as corporations find "green marketing" suddenly profitable. As global membership has exploded, new offices have been opening

194

pell-mell: Norway, Chile, Greece, Ireland, Moscow, Kiev, Rio, São Paulo, Costa Rica, Japan—all those just since 1988.

Greenpeace has also been negotiating with the television networks about "green programming," into which the international office is considering pumping several million dollars. McAllister envisions extending the organization's message into media ranging from comics and video arcades to investigative and mass-entertainment television, "doing something on the environment as successfully as *M*A*S*H* did with war."

Hollywood is already at work on its version of Greenpeace. Carolco Pictures is making *Warriors of the Rainbow*, with John Briley (*Gandhi*) as scriptwriter and Renny Harlin (*Die Hard 2*) as director. Robert Kline, one of the movie's producers, promises "passion" and "action . . . against a moral tapestry," with a projected release sometime this fall. The movie is based on the formative years of Greenpeace, when strategies and values that have shaped the organization ever since were laid down. One thing those rough-and-ready early years did not prepare Greenpeace to handle was its becoming wildly successful.

The name that now translates into fund-raising magic was born over a kitchen table in Vancouver, British Columbia. In 1970 two American families, alienated by U.S. Vietnam policies, joined with a Canadian law student in Vancouver to form the Don't Make a Wave Committee. They were protesting the upcoming U.S. nuclear-bomb test at Amchitka Island, in the Aleutian chain. Taking a cue from attempts by a Quaker protest ship, the *Golden Rule*, to sail into the bomb-test area at Bikini Atoll in 1958, the committee announced it would send a ship to blockade the Amchitka test.

Perhaps the history of the modern environmental movement would have been different if the name of the old halibut seiner chartered for the voyage twenty years ago this September had not been changed. She was originally the *Phyllis Cormack*. But shortly before leaving, Bill Darnell, a young Canadian social worker, managed to blend ecology and ban-the-bomb sentiments as neatly as anyone ever has: *Greenpeace*, they would rename her, and by 1973 the Don't Make a Wave Committee (so named to conjure the image of the bomb's unleashing tidal waves) would legally become the Greenpeace Foundation.

The nucleus of people who formed around this earliest manifestation of Greenpeace represented a stew pot of sometimes conflicting lifestyles and causes. Irving Stowe, a New England Jew converted to Quakerism, was opposed to the Vietnam War. He introduced the group to the Quaker concept of "bearing witness," putting oneself in the path of an objectionable activity. Jim Bohlen, a former Polaris-missile engineer from Bucks County, Pennsylvania, had come to Canada so that his stepson would not be drafted. Both men were members of the Sierra Club in those days but had grown frustrated with that group's lack of a stand on U.S. military policy.

Bohlen would leave Greenpeace in 1974, after Stowe, a vegetarian and health food devotee, died of stomach cancer and after what Bohlen calls "the hippy-dippy element" began to take over the leadership.

Enter Bob Hunter, a Canadian journalist and mover and shaker on the original Amchitka protest, who became president of Greenpeace during its second incarnation. Hunter's version of the group's early voyages is delightfully if impressionistically chronicled in his book *Warriors of the Rainbow*, on which the forthcoming movie is based. It begins: "This is the story of the first seven years of a movement that attempted to fulfill an ancient North American Indian prophecy of an age when the different races and nationalities would band together to defend the earth from her enemies."

A couple days out on the Amchitka voyage, Hunter passed along to the rest of the crew a pamphlet given him by a Jewish dulcimer maker who predicted it would reveal a path that would affect his life. It contained a Cree Indian prophecy made centuries before by an old grandmother, Eyes of Fire. It saw a time coming when "birds would fall out of the skies, fish be poisoned in streams . . . deer drop in their tracks and seas be blackened" by the white man's greed and technology.

The Indians, the legend said, would all but lose their spirit, only to find it again in teaching the white man how to revere Mother Earth. They would use the symbol of the rainbow to band together all the races of the world. They would go forth, warriors of the rainbow, to bring an end to the destruction and desecration. The legend lives on today in the form of Greenpeace's flagship, the *Rainbow Warrior.*

If Hunter could be mystical, he was also a thoroughly pragmatic journalist. He had a personal theory of "mind bombs," cobbled from the radical communications theories popularized by his fellow Canadian Marshall McLuhan during the 1960s. McLuhan, with his "the medium is the message" and "global village" concepts, preached that with modern communications, ideas could be moved out of the ivory tower and into the control tower with a speed and scope never before possible.

So it was that in 1971, Hunter clearly saw the little *Greenpeace* ship's sailing into the nuclear-test zone as an icon, a mind bomb whose physical presence, though relatively insignificant, had the potential to explode nuclear mind-sets in millions of heads around the world. He is frank about it in his book:

"We were engaged in a propaganda war. . . . Among ourselves we had always understood that Amchitka was not going to bring an end to the world [i.e., giant tidal waves and earthquakes that the press at the time was writing about]. But we had nothing more than images to hurl against the Atomic Energy Commission, so we threw the heaviest, most horrifying images we could."

In the strictest sense, Greenpeace's debut was a failure. Its members never made it to Amchitka, and the bomb was successfully detonated. But a major mind bomb had been exploded, via the wealth of international press coverage attending the group's voyage. Nuclear protests were in wide swing throughout America, inspired by many organizations as well as Greenpeace, and several months later the United States announced the end of tests at Amchitka "for political and other reasons."

None of this translated into tangible assets for Greenpeace, however. At the end of 1974, Hunter would write: "The resources of the Greenpeace Foundation came down to one thing: A brand name, if you like. . . . Our marketable, zippy, neat, easy-to-remember name . . . was all we had."

That next year, Greenpeace would detonate the mind-bomb equivalent of a nuclear blast, returning from the Pacific with a video showing Russians harpooning a sperm whale, narrowly missing a Greenpeace Zodiac running between hunter and quarry.

By then, Greenpeace was already a different organization from the one that made the Amchitka voyage. The antiwar, political ideology had been replaced by a near-mystical crusade to commune with and protect the only other creatures on the planet with large, highly developed brains. This was a crew that sometimes threw the I Ching, that believed in the laws of karma, that was partial to Quaaludes and that included a flutist to serenade their cetacean brothers. On the other hand they used good, old-fashioned research, including some neat espionage in the files of the International Whaling Commission, to ascertain where the Russians would be whaling. And Hunter's media instincts were keen as ever. The ship bristled with newsmen and camera gear.

Even so, at the moment of truth, it seemed as if some greater power was guiding the little ship. David Garrick, now a legislative assistant to a leading member of Canada's parliament, was a crew member then and kept careful diaries. At the time his hair was very long, and he called himself Walrus Oakenbough. He recalls the confrontation that would overnight make Greenpeace a household word this way:

"We had been on the campaign three months and shot a tremendous amount of film, and we had come nowhere near any whaling. The only reason we had a few minutes of film left was that the battery had died on the movie camera, or possibly it was a jammed trigger. Just at the moment when the Russian was preparing to fire . . . I saw [the cameraman's] finger tighten on the trigger of our broken camera. The harpoon fired and the camera started, and that gave us a place in history. There's no doubt that strange things happened out there."

"Soon images would be going out into hundreds of millions of minds around the world," Hunter wrote, "a completely new set of basic images about whaling."

Greenpeace was scarcely the only group trying to stop international whaling. At least twenty other organizations existed, and many, like Project Jonah, had been working diligently and effectively long before the upstarts' transcendent mind bomb hit the evening news with Walter Cronkite. Today it would make an interesting trivia question: Name another group besides Greenpeace that works to save the whales.

Nowadays, Greenpeace is richer, more powerful, more sophisticated and smarter scientifically than anyone dreamed even a few years ago. Revenues of $300 million worldwide by century's end are projected. Its membership profiles show higher numbers of young people than virtually any other major environmental group. The future seems unlimited. Yet it is fair to ask how much growth an organization that trades on activism and cultivates a "lean and mean" image can stand without losing something important.

Recently, Steve Sawyer, executive director of Greenpeace International, was paid a visit by a radical young environmentalist named Sam LaBudde—a type who might have sailed with Bob Hunter after the Russian whalers in the seventies, or with Steve McAllister in the early eighties after toxic-waste dumpers on the *Aleyka*. But LaBudde had come all the way from California to give Greenpeace hell.

In early 1988, LaBudde smuggled a videocam aboard a Panamanian tuna boat and shot striking first-time footage of how dolphins were still being slaughtered by the netters. Dolphins school with tuna, and since the 1960s fishermen have followed them to find their quarry. The practice had diminished since the landmark U.S. Marine Mammal Protection Act passed in 1972, but the film confirmed that cheating still went on.

Earth Island Institute, a small San Francisco–based environmental group that now employs LaBudde, says Greenpeace abused an arrangement to share the video with them by claiming in its direct-mail copy that LaBudde was associated with Greenpeace.

What made such an action particularly galling, says Dave Phillips, one of Earth Island's executive directors, was that at the same time, Greenpeace was frustrating the boycott of the major tuna companies by refusing to endorse it. Then, when the largest U.S. tuna canners suddenly agreed to market only tuna guaranteed "dolphin safe"—a stunning victory for the activists—"Greenpeace shows up with about eight people at the press conference, portraying it as a great victory and handing out copies of Sam's footage with their byline on it."

Following that, it appeared that one of the big canners, Bumble Bee, was not complying with the dolphin-safe pledge, Phillips says, "and at a meeting of the whole dolphin-conservation community, Greenpeace refused to sign a resolution passed by twenty-three other organizations condemning Bumble Bee."

Herb Gunther, whose Public Media Center, in San Francisco, used to do some of Greenpeace's most effective ads ("Kiss this baby goodbye" for the big-eyed, snow white harp seal pups), calls its handling of the tuna-dolphin issue "reprehensible."

"Here is the group that is virtually synonymous with protecting marine mammals," says Gunther, "and they are being very slick, not actually saying they are for the boycott but implying it; and unless you were an insider, you wouldn't know how Earth Island fought tooth and nail on $12,000-a-year salaries, how David Brower [its founder] mortgaged his house and how Greenpeace, with its huge resources, actually opposed it."

Indeed, Greenpeace is not above using the tuna boycott to advantage. An example is this pitch used occasionally by some of Greenpeace's door-to-door fund-raisers: "Do you eat tuna?" (Yes.) "Well, you know the little 'dolphin-safe' label on the cans now? That's *the kind of thing* we work for."

Greenpeace says from its viewpoint, there is no hypocrisy involved in soft-pedaling the boycott. Unlike Earth Island and other groups based solely in the United States, it has been working hard to expand into Latin America. Tuna exported from Latin America is a big and growing source of precious hard currency, but those countries lag behind their U.S. competitors in the technology and enforcement to assure dolphin-safe tuna.

Mention of the issue brings Steve Sawyer to his feet, hand slapping a desk top as he talks. Sawyer, thirty-four, has spent a third of his life with Greenpeace and has played a key role in its evolution, from the days when he split his $1,000-a-month paycheck with seventeen unsalaried Greenpeacers to the present position of prosperity.

"The U.S. has been fucking up the world for decades, telling the banana republics what's good for them," Sawyer says. "That's sanctimonious, hypocritical behavior for a nation that consumes twenty-five percent of the world's resources with only six percent of its people. The tuna [boycott] was a clear case of the U.S. trying to chop off Mexico's or Venezuela's economic balls . . . further widening the gulf with Latin and South America. Yes, life is more complicated when you look at the whole world, but that's what Greenpeace has to do now."

Such complexities, which grow ever greater as Greenpeace pursues expansion into more parts of the world, can cause difficulties for a group that has made its rep on moving fast, hitting hard and reducing complicated environmental issues to good guy versus bad guy.

"I still admire Greenpeace," says Earth Island's Dave Phillips. "From a philosophical standpoint they are still activist, not conciliatory like most of the big environmental groups. But from an operational point of view they definitely have gotten so big they have lost their edge."

That edge, or image as a fast-acting, hard-hitting group, is a source of more than simply pride to Greenpeace. It is at the very heart of its ability to raise money and attract members faster than virtually any group that ever set out to save the planet (only the Nature Conservancy, which preserves land by buying it or by accepting donations from wealthy people and corporations, has a budget comparable to Greenpeace's worldwide).

"People support us because we take action, we give them a feeling they are not powerless," says Vickey Monrean, a member of Greenpeace's international board of directors and the guiding force behind a direct-mail operation that brings in about sixty percent of the revenues for Greenpeace USA. Greenpeace organizes and videotapes "focus groups" to keep abreast of the mood of its public. When asked to complete the sentence "Greenpeace is the organization that . . . ," respondents typically wrote the following: "the organization that gets jailed a lot; that gets captured by the Russians; blown up by the French; shot at by the Japanese; arrested by the Spanish; fights head-to-head; gets it done." In another survey, people were asked to describe what the headquarters of various leading U.S. environmental groups looked like. Greenpeace loved the person who said its probably resembled a lean-to.

Only a few years ago, that wouldn't have been far off the mark. Greenpeace USA inhabited a shabby walk-up over a drugstore off Du-Pont Circle in the nation's capital. Nowadays it rents several floors of a building in a modest neighborhood of Northwest Washington—no corporate law office, but it has plate-glass doors, security, elevators, modern decor and photo-video facilities that many a big news organization would envy (and these are only the backup for Greenpeace's main visual-media center in London).

Peter Dykstra, longtime media chief of Greenpeace USA, is at work here on his latest project to embarrass government officials in high places. He has tracked down the solar panels for the roof of the White House, dedicated by Jimmy Carter in 1979 and removed by Ronald Reagan in 1986. They are stored, in good condition, at a government warehouse, and Dykstra very much wants to liberate them and install them on the roof of the late Mitch Snyder's downtown shelter for the homeless. To drive home the bankruptcy of the Bush administration's energy policy, Greenpeace would simultaneously distribute a video from the Carter press conference. Amid a year in which American boys fought for Kuwait's oil fields, it is haunting to watch the video and hear Carter's words: "A generation from now, these [panels] can be a curiosity, a museum piece, an example of a road not taken; or they can be a symbol of a more self-reliant and secure country."

Creative stuff indeed, this caper. The success of such stunts, explains a veteran Greenpeace campaigner, "is not whether they immediately stop the evil—they seldom do. Success comes in reducing a complex set of issues to symbols that break people's comfortable equilibrium, get them asking whether there are better ways to do things." Yet there are limits to this: Many within Greenpeace feel that the organization has reached the point where it presently has so many actions going on around the globe that it is overwhelming the attention span of both the media and the public.

Outside magazine, in its recent consumer's guide to the top twenty-five environmental groups, gave Greenpeace USA a mere 4 rating on its "milquetoast/bomb-thrower index," which ranked every group's degree of activism from 1 (milquetoasts) to 5 (bomb throwers). Eight of the other twenty-five groups also rated a 4, including the Natural Resources Defense Council, Sierra Club Legal Defense Fund, Environmental Action and Earth Island Institute.

If Greenpeace is no longer defining the radical edge of environmentalism, then who is, and what do they think of the organization that started it all?

Paul Watson is founder of the Sea Shepherd Conservation Society, a band of adventurers with 15,000 members and two ships that the group has used to ram whalers on the high seas, destroy eighty miles of Japanese drift nets, sink half of Iceland's whaling fleet and blockade a Canadian fishing community with the threat of sinking any vessel that left the harbor to hunt seals. While "the Sheps" have no problem going far beyond Greenpeace's prohibition on property destruction, they draw the line at firearms, explosives and deliberately injuring humans. Their flagship is armed with a bank of water cannons, hooked to barrels of government-surplus lemon-pie filling—the slippery stuff is cheap as hell and great for repelling boarders, they say.

Watson, who was kicked out of Greenpeace in March of 1977 for striking back against a Canadian seal hunter (he knocked a club from his hand), is today embittered and almost presents a burlesque of the radical environmentalists' critiques of Greenpeace. They are, he says, "totally fraudulent . . . gutless bureaucrats . . . fleas waiting to pounce on the latest dog."

"I wish Paul would get some fresh lines," Dykstra fires back. "They say he dresses up in uniforms these days and makes the crew call him admiral."

Shepherd president Benjamin White is more thoughtful. He cites a recent hearing on pollution by a big pulp-and-paper company where both organizations were involved. "Greenpeace made a fine scientific presentation on the effects of dioxin in their discharge. . . . For our part we brought an 'aperitif'—a container of the plant's discharge—and Paul offered county officials a drink. We were outrageous, and we were nearly thrown out. Between us and Greenpeace, I think we got the message across, but you see the difference."

Dave Foreman, founder of the radical group Earth First! (a high 5 on the bomb-thrower index) and currently fighting federal charges that he tried to destroy an energy facility, says: "The older I get, the more tolerant I get of different approaches. I just don't think anyone's got *the* answer, and we need all the approaches we can get."

If he has concerns about Greenpeace, Foreman says, it is that "fund-raising on the scale they are doing it can become the tail that wags the dog." (Its direct-mail fund-raising, for example, requires that forty-eight cents of every dollar raised be plowed back into sending out millions of pieces of junk mail. Greenpeace's Vickey Monrean says such mail has substantial educational value and is not merely junk.)

For an organization so committed to changes in society, Greenpeace still ranks poorly, along with the rest of the big environmental groups, in attracting minorities. Women have done much better there in recent years, however. "I think you could say we've come a long way, baby," says Monrean, who came to Greenpeace from a job as assistant to the president of NOW. Women, she says, currently coordinate three of the four major campaigns at the organization's international level and hold three of the five positions on its international board. She feels women may turn out to be best suited to running what many of the organization's leaders see as the Greenpeace of the future, expanding increasingly into regions and cultures where its traditional confrontational style is less at a premium than negotiation and diplomacy.

For all its critics and growing pains, Greenpeace scarcely seems ready to succumb to terminal establishmentarianism. Potential targets take the group very seriously. Ketchum Communications, a national advertising and public-relations firm, recently prepared a model "crisis-management plan" for the Clorox Company, detailing "worst-case" environmental scenarios the giant corporation might face. Actions by Greenpeace dominated the list of nightmares. Greenpeace has been running a worldwide campaign to end use of chlorine and its derivatives in the pulp-and-paper industry by 1993. Chlorine, compounds of which are used by Clorox in its bleach, is the base chemical in an array of pesticides and other toxins that are polluting waterways, according to Greenpeace. The document notes that the Greenpeace campaign, especially in Europe, "has been hitting home . . . and public response has begun to show up in lower use of household chlorine bleach in some areas."

Greenpeace remains proud that in the United States it has never been invited to join the so-called Group of Ten—the loose coalition of mainstream U.S. environmental organizations like the Sierra Club and the National Audubon Society. They do uncooperative things, like refusing to join the popular call for a boycott of Exxon after the *Valdez* spill in Alaska. The boycott ignored the real source of oil spills—our addiction to oil, Greenpeace said. Its alternative was to run national print ads featuring a mug shot of Joseph Hazelwood, captain of the *Valdez*: "It wasn't his driving that caused the Alaskan oil spill. It was yours."

And Greenpeace was virtually alone among major environmental groups (Friends of the Earth was the other) to quickly stake out a strong and well-argued position against the Gulf War, a move that probably cost it financial support. "For us it was relatively straightforward," says Peter Bahouth, the thirty-eight-year-old lawyer who is executive director of Greenpeace USA. "The war was going to fundamentally change this country. A big victory meant a boost in militarism, sucking off resources from social needs, promoting more nuclear weapons and increasing a push to continue the oil-based economy that is at the root of so many environmental problems, and it reaffirmed violence as the way to resolve conflict. If you were primarily concerned about the preservation of the planet, opposing the war didn't take much of a leap."

Greenpeace says it has actually bolstered its strike-force capabilities by raising responsibility for "direct actions" to the level of a separate department, headquartered in Washington. "We have skills now from mass spec analysis [of toxic chemicals] to high-altitude parachuting to electronic, nonviolent warfare," says Howard "Twilly" Cannon, the department's director. He runs summer camps to teach skills ranging from Zodiac operation and safety to climbing smokestacks.

Cannon says it is direct actions like the one Greenpeace mounted recently against Japanese drift netters on the North Pacific that show "the power that this organization can have now." The Japanese, he explains, are setting 20,000 to 50,000 miles of net *each night*. "Seeing their fleet set nets was like watching combines roll across Kansas," says Cannon. Drift nets are indiscriminate harvest gear, entangling everything that touches them, from squids to sharks to dolphins and sea birds. "It's how you squeeze the last little bits of life out of the ocean," Cannon continues. The Greenpeace action, when translated to the nightly news, appeared bold and spontaneous, but in fact it was preceded by eighteen months of meticulous research into the habits and methods of the Japanese fishing fleet, supplemented by electronic tracking of its movements.

Greenpeace employed divers and cameramen using special underwater lighting apparatus to obtain scientifically meaningful evidence of what the nets were actually catching per mile—evidence that would later be presented to the United Nations in a call to ban the netting. Ever mindful of providing photo opportunities along with the facts, Greenpeace took a few miles of Japanese drift net home and stretched it out on the Mall in Washington (the net later was shipped back to Japan). It was all a long way removed from Steve McAllister and his crew's windup Godzilla of less than a decade before. The cost of the whole campaign, about three-quarters of a million dollars, would have amounted to the whole Greenpeace USA budget a decade ago.

Cannon shares his office with Mike Roselle, cofounder of Earth First!, who came to Greenpeace in 1986. Roselle remains the uncompromising sort; he chose an extra sixty days in jail rather than agree to the terms of probation after a Greenpeace action in which activists scaled Mount Rushmore to protest acid rain.

"Greenpeace does take a long time to do stuff nowadays, and one part of us or another is in perpetual crisis," Roselle says. "But Greenpeace is magic, too, and you just have to accept it warts and all or you'll burn out." So far he thinks the organization has avoided becoming more conservative as it has grown. "I don't campaign fundamentally differently here than I would have with Earth First!, and I didn't have to promise Peter Bahouth I wouldn't spike trees on my off time."

But, Roselle adds, "I would work for the National Wildlife Federation [among the most conservative of environmental groups] if they had the right campaign."

In saying this, Roselle suggests something as essential to Greenpeace's success as its media appeal. Greenpeacers have a quality of being at one and the same time wildly idealistic and also exceedingly pragmatic. For example, the concept of bearing witness, a religious tradition often associated with Greenpeace and important as an end in itself to the group's Quaker founders, is seen by most leaders there now as an effective tactic, nothing deeper. Indeed, it is not unusual within the organization to hear people characterize Greenpeace itself as simply the most effective means to an end. "I'm no environmental purist," says Steve McAllister, who once took time off from Greenpeace to make a modest fortune in real-estate development. "Greenpeace's strength, its attraction to me, is that it cuts across all these lines—poverty, politics, economics, social conditions. Society is like a gigantic blindworm, and various forces prod it this way and that. I always saw Greenpeace as one of the best sharp sticks to prod it." The bottom line is that the only real Greenpeace credo, within broad limits such as nonviolence, seems to be "whatever works."

And whatever works is constantly being reassessed. "A colossal experiment," Bob Hunter once described the early Greenpeace in the seventies. A decade later, Peter Dykstra was characterizing Greenpeace activities as "an experiment, frankly. . . . We have no idea what we will be doing in five years, and like it that way." And in 1991, talking with Peter Bahouth, it is not long before he tells you of the group's struggle to avoid bureaucracy: "We're running a big experiment."

Where in fact is the great experiment that is Greenpeace leading? Answers to the organization's future, most observers say, lie as much as anywhere with a graying, media-shy Canadian who lives in Rome and counts himself a friend of Mikhail Gorbachev. David McTaggart, or McT, has bridged every era of Greenpeace, resurrecting it from vicious infighting after Hunter and others left the scene in the late 1970s. He remains at once the most respected and most controversial figure within all of Greenpeace, and his vision, more than any other individual's, guides the group to this day.

It is not long into an interview with McTaggart before the man who is chairman of the board of the world's largest and most far-flung environmental organization states a few firmly held personal principles: "I don't like meetings. I don't like organizations. I don't like large groups. I don't like consensus."

That seems incongruous at the least, given Greenpeace's phenomenal growth in the last decade, an interviewer ventures. "For me it was a cold-blooded decision," McTaggart says. Pre-Greenpeace, he worked alone and preferred it that way, both as a world-class singles badminton champion and later as a California builder who made and lost a considerable fortune. "But I learned you either work by yourself or you get strong—in between you get squashed."

Get big and get strong, or get squashed—he will restate that theme again over a couple days of conversations. It is a credo forged between the massive steel hulls of French warships that played a cruel cat-and-mouse game with McTaggart during his first Greenpeace protest two decades ago, threatening to crush his thirty-eight-foot wooden sailboat between them like an eggshell, 3,000 miles out in the South Pacific.

Answering an ad in a New Zealand newspaper changed his life. The Greenpeace Foundation was looking for anyone with a boat that could make the 7,000-mile round trip to Moruroa Atoll to put themselves in the path of the French nuclear bomb to be exploded there. McTaggart had the boat, and he had the time. After a tragic explosion devastated one of his construction projects and badly injured two employees, he lost his taste for the high-rolling business lifestyle and put what was left of his assets into the wooden ketch *Vega*, planning to spend the rest of his life as a wanderer in the South Seas.

He was no activist, nor had he ever heard of Greenpeace. Still, it was located in his hometown, Vancouver; or perhaps it was destiny, the old Greenpeace karma working. He took the job. No one else to McTaggart's knowledge ever applied. Bearing witness, Greenpeace style, to sins against the planet did not motivate him so much as French arrogance in illegally cordoning off from all boating 100,000 square miles of international waters around the bomb site. "That really pissed me off, because the seas represent the only true freedom left," McTaggart says.

He and *Vega*, renamed *Greenpeace III*, would make two voyages into the maw of *la bombe*, infuriating the French with the costly delays and international publicity the protests generated. The first trip was not auspicious. *Vega* was deliberately rammed by the French navy after days of playing a game of high-seas chicken. The French were able to explode their bomb when high winds briefly blew *Vega* out of the test zone. From isolated Moruroa, they were able to control news of the protest so that by the time McTaggart and crew limped back to New Zealand, even Greenpeace in Vancouver felt the brave expedition largely a failure.

McTaggart, exhausted and dispirited, nonetheless set about raising funds for a return to Moruroa. "For a few days," he would later write, "I had known that the tests were being delayed because we stood in the path of the fallout. I had known then that as surely as men possessed the power to build the bomb, they possessed the power to stop it from being built. . . . And having *known* that, for however brief a time, it was not something I could lightly put away. Perhaps it was something I could not put away at all."

The next summer the French got rough. They would teach McTaggart and his crew of three a lesson Greenpeace would never forget. As *Vega* bobbed in the test zone, they saw a strange, small craft (the inflatable Zodiac boats later adopted by Greenpeace) skimming at incredible speed across the waves, with several muscular French commandos aboard. McTaggart, in the book he did with Bob Hunter (*Greenpeace III, Journey Into the Bomb*), describes what happened after the boat was boarded and he was pinioned:

"The truncheons were flailing, each blow rattling my teeth so that it seemed they would be shattered and my spine and ribs and skull would cave in. Back. Neck. Head. Kidneys." And then: "Something crashed into my right eye with such incredible force that it seemed to come right into the middle of my brain in an explosion so that I thought that half my head had been torn off. . . . When I came to, half of my head seemed to be a vast hole. Only one eye would open. [Feeling for the other] I realized there was nothing there, just a wet pulpy socket."

Seated now beneath pictures of the beating that decorate his otherwise spare Rome office, McTaggart genially offers coffee and talks dispassionately about the incident of nearly two decades ago: "They were very professional. They even made their own truncheons from black copper-core coaxial cable so it wouldn't leave a mark. The eye was a mistake— one of them let himself get mad for just a second." McTaggart's girlfriend, Ann-Marie Horne, managed to click off a few blurred, grainy photos of the commandos, film she hid in her vagina during a subsequent French search of the boat. The resultant publicity embarrassed France around the world. The French, as Bob Hunter might have put it, had mind-bombed themselves. Later that year, in November 1973, the government announced the phaseout of all above-ground nuclear testing.

Doctors restored McTaggart's eye to its normal appearance, though the optic nerve was permanently damaged and he stands an excellent chance of developing glaucoma. Perhaps the French hoped the incident would end with their phaseout of the atmospheric tests. But they never knew McTaggart the world-class badminton player, who even today becomes excited when he talks of sprinting eleven miles during an average match and "watching your opponent's eyes—when the oxygen starts shutting down to his brain, you see that slight glaze in the eyes, and that is when you go for the kill. Every game you ever play is so hard, you never want to play again."

But you keep playing. McTaggart, nearly broke, pressed on to the enemy's court. He took a three-dollar-a-day room in Paris, where he would spend the next few years, his health and personal life unraveling, doggedly pressing charges against the French navy. He eventually would win on the minor charge that *Vega* was illegally rammed, but his charge of piracy, his own lawyer advised, was a hopeless quest. For that the French government would virtually have to admit its own guilt. In June 1976 the piracy case was thrown out as predicted, but in reading the verdict, the court official, speaking for the French government, brought gasps from the courtroom as he admitted that "McTaggart may have helped to persuade the French government to decide to choose underground tests."

"I knew I had never played a game so well," McTaggart would later write.

In the years that followed, McTaggart almost single-handedly gathered under the umbrella of an international leadership the splintered, warring factions that then passed for Greenpeace. "I don't consider that Greenpeace as an organization existed before that," McTaggart says. "Before it was just a lot of groups around the world using the name."

Since then, "he's been the most influential single individual in the organization by several orders of magnitude," says Steve Sawyer, expressing a view widely shared inside and outside Greenpeace. "He has incredible drive, incredible charm, incredible intuition, and when things get between him and an objective, he can be just implacable."

This does not, however, always translate to popularity, Greenpeacers are quick to note. "I have enormous respect for McTaggart, but he and I have never gotten along," says one longtime official who was among many disillusioned by the chairman's decision to move the group into Russia in a major way.

By all accounts, McTaggart got high on the stock of Mikhail Gorbachev in 1984, at least six months before even the savviest international political pundits in the United States realized he was not your average Russian bear. McTaggart has since traveled to the Soviet Union thirty-eight times. Greenpeace is legally approved in the Ukraine to bring people in and out and to hire and buy vehicles and property. It raised a lot of money for itself and for joint environmental projects with the Soviets in 1989 with the release of a benefit rock album with twenty-four leading Western artists and bands. The organization has opened a research lab and a clinic in the Chernobyl area to help radiation victims.

"We could give lessons to a lot of multinationals on how to operate in the Soviet Union," says McTaggart. "I cannot believe anyone can work with the environment in this world and ignore a place with resources and problems the size of the Soviet Union's."

Russia is just part of his vision for turning Greenpeace into a truly global force, changing its nature from a North American–European–

dominated group to one that is also at home in Latin America, Asia and Africa. The confrontational tactics that have come to symbolize Greenpeace, McTaggart says, will probably not be the ones that work as it expands into different cultures and governmental systems that are often far less tolerant of protest. "You have to respect the way each country operates, and dead martyrs aren't worth a shit anyway," he says.

As the experience with the tuna-dolphin boycott illustrates, the road to such a future can be bumpy. And the Russian expansion, which once seemed crazy, then seemed an act of genius, now [1991] has the potential to be so much wasted effort as the U.S.S.R. unravels and Gorbachev grows embattled.

For McTaggart, mention of such risk-taking gets right to the heart of what Greenpeace must be about. "Look, we were never made to be a bureaucracy," he says. "We're made to be expendable, and that is a fucking fact. I hope we're big enough and strong enough now to take on some issues big enough to maybe shut us down, or at least a part of us.

"People say $160 million a year [projected worldwide revenues for Greenpeace] is a lot of money," McTaggart continues. "I say bullshit. Compare it to the interests we're going up against—militaries of large nations, multinationals, nuclear interests. . . . They spend in an hour what we do in a year.

"Not many people understand how we really work," McTaggart says. "We would always prefer *not* to do actions, because of the money and the risk. We try to get the situation clear in our own heads first, do the science, then have quiet conversations with lots of people, from companies to presidents of countries. If they do whatever, that's all you'll hear about it. If nothing happens, then you go to the action.

"Antarctica was a perfect example of quiet activism," McTaggart continues. "Setting up our base down there was like a dog peeing, staking out territory. If anyone had really had a legitimate claim to Antarctica, they would have run us out, and they didn't. In the long term, saving that whole continent will be one of the major things Greenpeace did, yet I don't even know how you'd write about all the behind-the-scenes work we did.

"You do not undertake any campaign where you cannot break the issues down into very simple areas anyone can understand," says McTaggart. "You take risks but you don't gamble foolishly. We have always, on purpose, looked rough and haphazard, but most of it's been carefully thought out and planned, sometimes years before. In doing actions, I always want to pick targets where I feel we have at least a three-to-one odds [of succeeding]."

McTaggart is reluctant to quote odds on the largest campaign of all, determining the future of Greenpeace: "That is the struggle, isn't it? To be large and still act quickly, and be semidemocratic. You can't be an up-front organization and get consensus all the time. I can't find any organization, any religion that has done it."

The other tough hurdle, McTaggart says, is "the mistake I made in the beginning, to think people care more for the environment than for their various national interests. It is one of the hardest things to overcome . . . thinking that your country is the greatest; we are taught that from birth. I think Greenpeace recognizes that better than any other organization, but it is still hard finding enough good people with a truly international viewpoint."

Asked to describe his own underlying philosophy of the environment, McTaggart characteristically replies: "I'm not a philosopher, and I don't know if I really am an environmentalist. I have a lot of respect for the trees, the oceans . . . but the main thing I have to contribute is, I was in business before this, and I can understand the opposition, their bottom line, profit and loss."

Brian Fitzgerald, a Greenpeacer since 1982 who has been McTaggart's personal assistant for six years, says the chairman's strong identification with Gorbachev may be more than coincidence. "He's somewhat in Gorbachev's position right now," says Fitzgerald. "He's got a lot of Yeltsins. Even ideas like Antarctica, which everyone now likes, were pushed through at a political cost. He's the only one from the early generations of Greenpeace still in power, and that still carries a lot of weight; but even that's changing. Nowadays we're seeking people with qualities of training and professionalism that are very different from the up-through-the-ranks types we have always had. Probably the value of being an old-timer is diminishing. Part of the tension in Greenpeace with David right now is that some of the newer people see survival of the organization as more important than the issues. David's no crazy man, but he feels the risk-taking is vital."

Increasingly these days, Fitzgerald says, his boss will sit through another lengthy meeting and wonder, "Could we end up like the United Nations?"—that is, global to the point of paralysis.

That concern is voiced most pointedly by an outspoken McTaggart critic, Allan Thornton. Now chairman of the Environmental Investigation Agency, which has exposed illegal trade in ivory and endangered species, Thornton founded Greenpeace in Britain with McTaggart in the 1970s and remains on the boards of the organization in Canada and Britain. "David's point of view is that we have a sort of eco–Manifest Destiny to grow and organize into every part of the globe," Thornton says. "Mine is that what Greenpeace is best at is specific campaigns. That doesn't exclude us from anywhere in the world, but it might just mean sending a person in with a phone and a fax—not years of planning and organizing from the top down, like we're doing in Latin America and Russia."

It is early May, still cool back in Amsterdam but a warm, lazy afternoon at Peace and Plenty Farm, on a bluff of the coastal plain looking out toward the Potomac River near where it feeds into Chesapeake Bay. This is Steve McAllister's home place, purchased a few years ago with an ex-Greenpeacer. McAllister, who just last winter seemed brimful of visions for the future of Greenpeace, has abruptly quit his deputy director's job, and now, ambling along a red dirt lane among fields of waving clover, he is talking about settling down, raising organic fruit and aquaculturing catfish for a living.

He does not want to talk in specifics about his reasons for leaving, but yes, he was increasingly troubled by the direction he felt Greenpeace was headed. Colleagues in the organization say it would be stretching to interpret the resignation as symptomatic of Greenpeace starting to unravel. There are, after all, more than a thousand full-time employees now, and Mac already left once before for a couple years after losing an internal power struggle in the mid-1980s.

On the other hand, they say, don't dismiss this as insignificant; Mac did, to a great extent, embody the spirit of Greenpeace that so captured the public imagination during the last decade. That spirit is a hard thing to capture on paper, but one example is the raid he led in 1983 on a toxic-chemical recycling plant in Baltimore. Greenpeacers drove their inflatable boats right up the company's discharge canal, a little ditch flowing through its property into the Chesapeake Bay. It was tidal and therefore, in legal jargon, "navigable waters of the United States," which meant that as long as Greenpeacers stayed in the horribly polluted ditch, the company couldn't evict them and couldn't dump more wastes. They camped in the ditch for four days and nights, as public pressure to close the plant mounted (the state later shut it down).

Charges were brought against Greenpeace for destruction of property—a fence the company had erected across the ditch. The judge gave probation with a mild fine. As the protesters emerged from the courtroom, the city's chief prosecutor halted them. He had refused to try the case, he said (an assistant took it), and he stuffed a $10 contribution to Greenpeace into one of the activists' shirt pockets. What they were doing, he said, might be legally wrong, "but you are right. . . . That company is killing our bay."

Never mind that in the greater scheme of decades of striving by dozens of other environmental groups and government agencies to save Chesapeake Bay, Greenpeace's contribution has been minimal. It is that ability to outrage, to *inspire*, that is almost unique to Greenpeace among the environmental community. And if it were to lose that, it is doubtful in the long run whether any amount of dollars or members could compensate.

The tour of the organic apple orchard is over. Somehow, McAllister doesn't seem really excited about apples. But there is this other project that he has in mind that gets the juices flowing. Chesapeake Bay is well represented by environmental groups, but it seems to McAllister the place could use another one: nothing big; something lean and mean, beholden to no one, speaking with a bold, even outrageous voice, speaking just for the bay and willing to take risks. The more he thinks about it, the more he likes the idea. In fact, he says, it sounds like just the sort of thing Greenpeace could be talked into funneling some money into.

The Greening of the Golden Arches

Bill Gifford

*I*t looked like a plastics executive's dream come true. There stood Jackie Prince, a scientist with the Environmental Defense Fund [EDF]—the enemy—at a McDonald's stove, flipping Big Macs. To industry thinking, it was just where she and her kind belonged, forever.

But in the end, it was bad news for plastics. Prince and two other EDF staffers worked in a McDonald's for a day last summer as part of a joint EDF-McDonald's task force searching for ways the fast-food chain—the sacred ark of the throw-away society—could reduce its 2 million-pound-per-day Niagara of waste.

The first answer came in November [1990], when McDonald's announced it would deep-six its signature polystyrene clamshell sandwich boxes. Then in April [1991], McDonald's released the task force's sweeping 138-page report, which went well beyond clamshells. Among forty-odd other changes, the chain said it would recycle all its corrugated cardboard, use less paper in its napkins and test a refillable coffee mug.

In marketing terms, losing the clamshell paid off immediately. McDonald's swept a January [1991] *Advertising Age*/Gallup poll as the most environmentally responsible fast-food chain, in consumers' eyes, beating out Wendy's, Kentucky Fried Chicken and Burger King. But the clamshell decision—and the company's highly unorthodox alliance with the Environmental Defense Fund—didn't seem to please many others. "McDonald's Caves In," moaned *Forbes*, while the Naderite *Multinational Monitor* accused EDF of "Degrading Environmentalism." The plastics industry launched a well-funded but dull-witted ad campaign extolling the virtues of polystyrene and attacking the clamshell move as environmentally unsound.

The task force's proposals, which McDonald's says could cut its waste by three-fourths, will fundamentally change many of the company's operations. But the plan's impact is being felt far beyond McDonald's dumpsters. As a blueprint of a new way for corporations and environmentalists to deal with one another, it testifies to the new economic clout of environmental groups.

The strange marriage between EDF and McDonald's dates to 1989, when the company's general counsel, Shelby Yastrow, was scheduled to appear opposite EDF's executive director, Fred Krupp, on a cable TV show. Yastrow called Dan Sprehe, a legislative analyst with the company, and told him to check out Krupp's group. "It's hard to believe," Sprehe says, "but nobody around here knew much about them two years ago."

Founded by Long Island scientists in 1967 to fight local DDT spraying, EDF's motto in the old days was "Sue the bastards!" But since Krupp assumed leadership in 1984, EDF has promoted a kind of environmentalism that tries to satisfy economic needs as well as ecological concerns. Spend more than a half hour with Krupp, a slightly nerdy but persuasive thirty-six-year-old, and he'll launch into a spiel about Solving the Big Problems—global warming, rain forest depletion, protecting Antarctica—by harnessing market forces and removing the incentives to plunder and pollute.

In an influential 1986 *Wall Street Journal* op-ed, Krupp called EDF's approach the third stage of environmentalism; it followed the first stage, exemplified by the early Sierra Club and Teddy Roosevelt–style conservationists, and the second stage of lawyers and lobbyists, who passed the sweeping environmental laws of the sixties and seventies. Environmentalists in the third wave, as it is now known, have grown frustrated with the legislative process. When powerful economic interests collide with a small, vocal interest group, the result is often blood-soaked compromise and bad law. Bureaucrats then must write the rules to enforce the law, but they sometimes don't get around to it for a while. When the new Clean Air Act was passed last year [1990], EPA rule writers were still dawdling over the previous version, passed in 1977.

Alongside the standard lawyers and PhDs, EDF's staff featured economists before economists were environmentally cool. The group ghostwrote President Bush's acid rain bill, which introduced an innovative market mechanism to control pollution, and is now developing market-based distribution systems for water in the West.

It isn't the kind of environmental group that McDonald's is used to bumping into. McDonald's had linked up once before with environmentalists, funding a booklet and a rain forest poster with the World Wildlife Fund (the Sierra Club, among other groups, had spurned advances by McDonald's). And on its own, McDonald's had announced its intention to spend $100 million annually on products made from recycled materials and helped start a polystyrene-recycling plant.

But all those efforts had amounted to naught in the marketplace. By the late eighties, more and more consumers were choosing products based on environmental criteria, and the McDonald's clamshell had become a symbol of ecological evil. Schoolchildren vilified its beloved mascot, calling him Ronald McToxic, and mailed their McTrash back to the company—personally addressed to Yastrow. Consumer pressure was also being expressed in the Styrofoam-packaging bans popping up all over the place. With landfills running out of room, especially in the Northeast, the actual cost of disposing of garbage was rising as fast as the company's PR bill. In addition, McDonald's was entering the worst domestic sales slump in its thirty-five-year history.

Something had to be done, but what? For environmental information, McDonald's usually turned to its suppliers—whose goal was to sell more packaging, not less. And its environmental affairs department consisted primarily of general counsel Yastrow and assorted flacks, who defended Styrofoam on the creative grounds that it "aerates the soil."

At Yastrow's invitation, Krupp jetted out to company headquarters, nestled among prairie wetlands (and ringed by a McNature Trail) in Oak Brook, Illinois, for a chat with Yastrow and Ed Rensi, the head of U.S. operations. "One thing I told them," says Krupp, "was that if McDonald's wanted to be a leader on the environment, they weren't going to get there by just giving money to environmental groups. The way to get there would be by changing their operations."

What Krupp saw in McDonald's was a golden opportunity to harness market forces on a grand scale. As the nation's largest consumer of packaging, McDonald's has market clout that approaches historic dimensions. In the sixties, McDonald's forced the dairy industry's wholesale shift from steel cans and glass bottles to plastic and paper cartons because founder Ray Kroc couldn't stand to waste the space between the round cans. And as Krupp well knew, McDonald's thrives on change. "Beneath [a] cloak of uniformity hides a corporate culture that worships flexibility," writes John F. Love in his thorough history of the company, *McDonald's: Behind the Arches.* "Overnight, [managers] drop carefully laid plans that have been torpedoed by a shifting market."

Over the spring and summer of 1990, McDonald's and EDF hashed out the agreement that set up the joint task force. EDF insisted on its right to criticize the company and forbade McDonald's to advertise the relationship (though EDF itself has been diligent in publicizing the task force). McDonald's required the three EDF task-force members to work in a restaurant. The company wouldn't be bound to adopt any of the task-force recommendations; issues such as rain forest destruction, global warming and the high-consumption, highly disposable nature of

McDonald's' business were off-limits. And either side could scuttle the project at any time.

Conditions set, the task force began work in August [1990], focusing first on the fate of the clamshell. Its seven members spent long meetings debating various new wrappers and boxes submitted by suppliers. They even considered an edible wrapper, which McDonald's operations director Keith Magnuson described as "a little chewy," before settling on a layered tissue-and-plastic wrapper with insulating air bubbles.

But in late October [1990], word reached EDF that a pro-clamshell faction within McDonald's was attempting an end run around the task force. The company was about to renew its commitment to polystyrene. Krupp called Rensi and put the project on the line. The task force has come up with an alternative, he said. If the announcement were made, EDF would publicly blast it. The next Monday, Krupp flew to Oak Brook and made his pitch to senior management. Two days later, McDonald's said good-bye to its beige, sky blue and pearly white boxes.

But the move to paper "wasn't much of a switch," says Jan Beyea, senior scientist with the National Audubon Society. "It sent the wrong message, that by switching to paper you're doing the environment a favor. Paper is made with tremendous chemical and industrial processes." The new wrapper, concedes Richard Denison, an EDF member of the task force, "is by no stretch of the imagination a recyclable material." And besides, McDonald's didn't fully kick the polystyrene habit: Breakfast entrees are still encased in bad old plastic [as of 1990].

The real reason McDonald's switched can be found in the company's 1990 annual report, which states: "Although some scientific studies indicate that foam packaging is environmentally sound, customers just didn't feel good about it." Or as Yastrow put it: "That clamshell package was the symbol that everyone glommed onto. We knew if we got rid of that thing, it would be like pulling forty thorns out of our paw."

But if the market force of consumer reaction provoked the dubious clamshell move, it also motivated more positive gains. The final plan, released in April [1991], contains three dozen or so initiatives, of which customers will notice only a handful. Ketchup packets will get larger so customers will use fewer. Carryout bags will be made of recycled, unbleached paper. Napkins will be shrunk by a fifth (but refolded to appear the same size).

Most of the changes will take place behind the counter. In part, that's by design: McDonald's wouldn't dare inconvenience its customers in the slightest way. But it's also because more than eighty percent of trash at a McDonald's comes from behind the counter, as the task force learned from a waste audit of franchises in Denver and Sycamore, Illinois. Bulk shipping containers and corrugated cardboard alone account for a third of the typical franchise's garbage output. So the task force decided to recycle all the corrugated—eliminating, in one stroke, nearly 350 tons per day of landfill-destined trash. Then, to create demand for recycled corrugated, McDonald's ordered its suppliers—the companies that sell everything from meat to coffee cups to McDonald's—to use boxes with a minimum of thirty-five-percent-recycled content.

McDonald's resisted the whole concept of reusables—plates and service—though EDF pushed hard for months. The chain rejected proposals for refillable cold cups and serving burgers on plates because customers tend to leave with their drinks and because hot-off-the-grill sandwiches go into the same wrapper for both take-out and eat-in customers. McDonald's did agree to test reusable shipping pallets and a refillable coffee mug.

EDF also nudged McDonald's into testing far-out waste-disposal techniques such as composting. Ten McDonald's are sending coffee grounds, eggshells and food-coated paper to a Maine composting plant, where organic waste decays naturally into dirt that's sold to farmers and landscapers. Such tests are just a tiny step toward dealing with food waste, which constitutes a third of the typical store's trash. Every day, the average McDonald's tosses out eighty-one pounds of unsold Big Macs, mushed fries and other perishable items, which adds up to a whopping 694,000 pounds of wasted food nationwide.

The task force ignored at least one other environmental disaster propagated by the McDonald's system: cows. Their grazing habits cause erosion, their dung seeps into groundwater and greenhouse gases pour from their nether regions. "The best thing McDonald's could do," says a scientist with a rival environmental group, "would be to get out of the business of marketing meat."

Many environmentalists seem to have difficulty condoning the existence of McDonald's. Few would be caught dead eating there. Greenpeace executive director Peter Bahouth says that if he were president of the company, he'd "shut the doors." But Krupp, who occasionally eats at McDonald's with his two children, doesn't see it as black or white, McDonald's or Mother Earth. EDF's brand of environmentalism takes what it can get and counts even small steps as progress.

"We're not ideologues on environmental issues," says Krupp. The myth blocking progress "is the notion that all environmentalists should be trapped in one narrow set of tactics." He says: "I think environmentalists would become more powerful, more forceful and achieve greater results if we deployed more tools in our tool kit. We should continue to aggressively lobby, aggressively litigate, aggressively criticize corporate malfeasance and promote stricter regulation. We also should be able to problem-solve with corporations."

But Krupp's willingness to talk with the capitalist enemy causes some environmentalists to curse him (usually off the record) as a kind of green Benedict Arnold, brown-nosing big business and the White House. Especially angry are grass-roots groups like the Citizen's Clearinghouse for Hazardous Waste [CCHW]—founded by Love Canal survivor Lois Gibbs—which feels Krupp is stealing the credit for winning the clamshell war it started.

"He says, 'All I did was call Ed Rensi, and that did it,' " Gibbs says, "but Ed did it because of three years of grass-roots work." Says Krupp, "EDF was one of many voices that McDonald's listened to."

The rift between EDF and CCHW reflects a wider division among the greens, between talkers and fighters. EDF's sit-down-and-chat methods don't work for everyone. Three years ago Yastrow met with CCHW to discuss the group's demands. Both sides say the meeting was a disaster. And Michael Fischer, who heads the graying-but-still-radical Sierra Club, doubts his group would be any good at third-waving: "The future of EDF may be in the third wave, but damnit, it's not the Sierra Club's future."

But the third-wave future is now for McDonald's and for its wide-ranging network of suppliers, who are scrambling to meet recycling rules dictated not by the federal government but by one of their biggest customers. For years, corporate executives have argued that environmental considerations were a luxury they couldn't afford. McDonald's—confronted with skyrocketing actual and PR costs of garbage, and nudged and advised by EDF—has adopted a strategy based on the opposite notion. Environmental action was something it could ill afford not to do.

EDF took no money from McDonald's. In the competitive world of environmental groups, however, results equal donors, and EDF's direct mail now boasts of slaying the clamshell. But donors are also customers, not only for McDonald's but for Safeway, Anheuser-Busch, Procter & Gamble and hundreds of other firms now feeling the pressure of green consumerism. Krupp loves to talk about harnessing market forces, but it's more than that. As the greening of McDonald's shows, Krupp and the movement he represents have become a market force.

PART VI

Working for Tomorrow

The essays preceding this section all document the inevitability of change: the environmental hazards posed by our modern age will ensure this. The true conflict lies not in whether change will take place, but rather what values will shape this change. Environmentalists are often criticized for tolling environmental Armageddon; yet many, in fact, dedicate their lives to creating new ways of doing things, and this section is dedicated to their vision of the future.

The possibilities for change are more varied, and more numerous, than many people think. One path takes us away from the twentieth century's obsession with technology. Howard Kohn explores new farming methods that might provide an alternative to America's chemically based agricultural industry and shows that we have choices in how and when we use the scientific advances of our times.

Yet if real change is to take place, many environmentalists think that a new framework to understand the world around us is needed. Perhaps no one is more knowing than David Brower of the moral and ethical choices that environmental issues create. Appreciated for his unyielding dedication, Brower is considered by many to be the soul of the modern environmental movement, and no collection of essays would be complete without his unique voice.

Part of Brower's message is that the future depends upon a reawakening of our personal connection with nature. Many Americans are trying to heighten their understanding and appreciation of the natural world. Don Henley,

E. L. Doctorow and Robert Bly give testament to their personal relationship with nature. These men speak specifically of Walden Pond, for many the spiritual home of the environmental movement in America. Their essays are selections from a book compiled to raise money to purchase the land around the pond in order to protect it from a massive development project. (As of February 1992, the Walden Woods Project has succeeded in purchasing one of the two parcels of land surrounding the pond; a parcel of land that will house an office park has not yet been protected.)

Personal responsibility seems to lie at the heart of the possibility for real change. And there have been signs that people are willing to alter their habits to protect the environment. The popularity of vegetarianism, eco-tourism, environmentally friendly products, and farmers' markets all point to our willingness to change.

Information *is* spreading about the effect of lifestyle on the environment. Yet the will to change requires more than a broadened supply of consumer products; it requires a rebuilding of our sense of responsibility toward the natural world, of our care and appreciation for it. Environmentalism has proven to be much more than a passing trend, and, hopefully, in the years to come the paths we choose to make will become clear.

Fields of Dreams: Old Farming Is New Again

Howard Kohn

*I*n 1981, the award-winning reporter Ward Sinclair was assigned to the farm beat at the *Washington Post*. After twenty years of general-assignment reporting, Sinclair figured this was no way to be treated. He was sure his byline would never again see the light of the front page.

But, in fact, Sinclair transformed his beat into one of the hottest on the paper, writing dozens of major articles about guys in overalls who didn't return his telephone calls till the sun went down. Sinclair was a tough and conscientious reporter. I remember him once at a news conference called by the Rodale Institute, the Pennsylvania research group that for the last half century has tried to turn back the clock on American agriculture. Robert Rodale, the son of the founder, was trying to impress us with the amazing success ("crop yields in the upper ten percent year after year") of a Lancaster County farmer who used neither pesticides nor petrochemical fertilizers.

"Yes, but is this someone who doesn't use chemicals because of his religious convictions?" Sinclair asked. This was the key question, because the farmer was, as a matter of fact, a member of the Amish sect. The Amish live by choice without most of the conveniences of modern society, including pesticides. Their self-denying way of life affords them the time and the discipline to put in long days of hard work, which is precisely what the use of chemicals has been designed to reduce for the great majority of American farmers. It seemed unfair to compare the religiously committed with everyone else.

Last year, to the astonishment of everyone at the *Post*, Sinclair quit his job. He was too young, at the age of fifty-five, for retirement, and he appeared to have nothing else lined up. One Sunday morning early this summer, I ran into him at a farmers' market in Takoma Park, Maryland, a suburb of Washington. A number of people in the news business do their shopping there, and it took a while before I realized that Sinclair was not buying any produce. He was *selling*. The lettuce, green beans and strawberries in his baskets were from his own farm in the Pennsylvania hill country.

Sinclair told me that he had sixty-five acres and that he had given his farm a most un-Amish name—the Flickerville Mountain Farm and Groundhog Ranch—lest anyone confuse it with a place of biblical injunction. He said he got up early, worked all day and fought bugs and weeds with no help from chemicals. Indeed, Sinclair has become in his new career as old-fashioned as any Amish man of the soil.

What's different is that Sinclair and many other farmers are standing the word *old-fashioned* on its head. After many years of low-key debate, there is intensifying alarm in the United States and Europe about the application of chemicals to growing food. New, highly publicized studies linking farm chemicals to increased rates of cancer, birth defects and other health problems have brought the issue into public view. The result is that "old farming" is now being called "new farming."

Sinclair left the big city for the sake of his sanity. "You get fed up with rush hour," he says. He didn't intend to be a crusader but now finds himself caught up in a national movement. Only a year ago Sinclair was interviewing state agricultural commissioners, such as Gus Schumacher of Massachusetts and Jim Nichols of Minnesota. Now the roles are reversed.

Schumacher and Nichols, along with Texas agricultural commissioner Jim Hightower, are among the country's most respected and forward-looking farm officials. All believe that the Chemical Age of American Agriculture is in decline. "Most American consumers will not eat food they believe has been poisoned," says Hightower, a leading candidate to become U.S. secretary of agriculture in the next Democratic administration.

The use of chemicals to ward off pests and boost production began in earnest after World War II, thanks to an advertising campaign by Dow Chemical, Monsanto and other chemical companies. This came to be known as the green revolution. Fears about possible health hazards from farm chemicals started to surface two decades ago. But not until this past spring did consumers begin to panic. Reports on 60 *Minutes* and in a *Newsweek* cover story about the heightened risk of cancer from apples sprayed with daminozide (trade name, Alar) did what earlier stories about pesticides had failed to do. Sales of apples dropped sharply, and so did those of other fruits and some vegetables. In New York, Los Angeles and Chicago, all apples were removed from school cafeterias.

Even before the frenzy of publicity about Alar, consumers had been getting nervous about the dangers of pesticides. Half the people interviewed by the Harris Poll last year said they would be willing to pay extra for food grown the way our forebears grew it. Many varieties of apples, for instance, are tasty and crisp without Alar. On his farm, Sinclair has an apple orchard consisting mostly of trees that haven't been popular since the nineteenth century—Winter Banana, Seek No Further, Arkansas Black and the one that, according to Sinclair, was Thomas Jefferson's favorite, Sheepnose. These and similar varieties are making a comeback across the country.

To be certified as a farmer who grows food free of pesticides has become good for business. A handful of major supermarket chains (Stop and Shop, Petrini's and Ralphs) have set up prominently placed, well-advertised display tables of certified produce. Meanwhile, an increasing number of states have programs identifying farmers who bring naturally grown products to market.

In June, with the wettest spring in memory sliding into summer, slugs as big as your thumb were laying waste to Sinclair's pepper patches. He had planted fourteen varieties, figuring this would be the best way to find which appealed the least to slugs. Then, too, he was expecting to discover which was worse: last year's killer drought or this year's killing rains.

Late one evening Sinclair was at his desk filling out an application form for a farmers' market in Rockville, Maryland. When he had started in farming, it hadn't occurred to him that supply might be a bigger problem than demand. But on his desk was an application for another market in upscale McLean, Virginia, and he already had a regular run to restaurants, delis, country stores and food co-ops.

Every Tuesday, Sinclair drove his truck into downtown Washington, parked outside the *Post* building on Fifteenth Street and carried inside special, hand-packed bags for 115 former colleagues who had become his customers. The executive editor, Ben Bradlee, was one, as were the cartoonist Herblock and the restaurant critic Phyllis Richman. While unloading his truck, Sinclair would often find himself approached by strangers desperate to get on his list. But Sinclair always turned them down. He felt that the point of a personal delivery was to have friends on the other side of the transaction.

With all his commitments, Sinclair decided to forgo the McLean market in favor of Rockville, which was offering $50 a day to all farmers—until recently an unheard-of idea among market owners. The antipesticide movement was building the confidence of Sinclair and other natural growers. "You have to be crazy to try to make it as an organic farmer, but I'm crazy enough to believe I can," says Sinclair, who expects to turn a profit this year. "I believe consumers will have their say."

What consumers have been saying for years is that they want cheap, plentiful food, which the green revolution provided. What they are saying now is that the food they eat must, above all, be safe. "The hidden costs to people's health were always there," Sinclair says. "We chose to look the other way."

In effect, consumers have become their own test subjects. Although DDT was banned from agricultural use a decade after Rachel Carson's book *Silent Spring* was published in 1962, scores of other pesticides eventually took its place, most of them without extensive testing by government regulators. In 1979 aldicarb, a cancer-causing pesticide, was found to have run off potato fields and contaminated drinking water on Long Island. In 1984 tons of grain had to be destroyed because of another pesticide, ethylene dibromide. In 1985 pesticides sprayed on watermelons in the western United States and Canada poisoned 500 people, many of whom required hospitalization. In 1986 the pesticide heptachlor contaminated milk and meat in seven states. In 1988 cucumbers with excessive aldicarb had to be confiscated.

These were all important news stories, some with Sinclair's byline. He began to wonder why modern farmers and consumers had accepted chemical helpers so unquestioningly. "Something is wrong when you can't eat what you grow," Sinclair says.

The next question was whether other methods of farming might succeed. Farmers who make do without chemicals must resort to such traditional methods as rotating crops, fertilizing with manure, planting a number of varieties, encouraging natural predators and—the most time-tested of all—wielding hoes and sickles.

"The longer I listened to the iconoclasts and the dissenters, the more sense they made," Sinclair says. "Bob Rodale, for instance."

Rodale [was] a restrained, gracious man, more of a visionary than a farmer, who inherited from his father, J. I. Rodale, a small publishing business and a mission. Despite suffering the slings and arrows of a skeptical press, the younger Rodale was able to turn the business into an empire (*Prevention* and *Organic Gardening* are his two most successful magazine titles) and the mission into a movement. He also had to overcome the relentless antagonism of the chemical industry, not to mention the unfortunate timing of his father's death (J. I. Rodale was singing the joys of organic living on *The Dick Cavett Show* when his heart gave out).

Until several years ago, Rodale had to support his small research institute in Emmaus, Pennsylvania, with private money, most of it from his own pocket. And yet Rodale was always feared by the industry—way out of proportion to the threat he posed. I happened to mention Rodale's name once to Robert Tennant, who was Dow Chemical's marketing man for pesticides. "Oh, so you've met the enemy," Tennant snapped.

Attitudes toward Rodale began to change in 1979, after the Carter administration sent USDA researchers to get his list of organic farmers for a study they were conducting about the economics of farming without chemicals. The USDA study, released a year later, found that contrary to conventional wisdom, organic farmers had as much chance as chemical farmers to make a profit. Unfortunately, the report was not filed until Carter's final days in the White House, and it was ignored by the Reagan administration.

Then, in 1982, a top Rodale official, John Haberern, came up with an idea. Why not circumvent the White House and push Congress to pay attention to the report's findings? Three years later, after exhaustive lobbying by Rodale and Haberern, Congress established the Low-Input Sustainable Agriculture (LISA) program. Under LISA the USDA was mandated to fund research into alternatives to chemical farming. Last year seventy-eight projects were funded, three at the Rodale Institute.

Few current federal programs are as radical. In the long term, LISA may well drastically reduce chemical farming. In the meantime, it gives political cover to elected farm officials such as Hightower, who must defend his belief in organic farming in a state dominated by agribusiness. Potentially, it will also provide Sinclair and other organic farmers the kind of federal research support that chemical farmers have enjoyed for years.

Of course, reports of the death of chemical farming can easily be exaggerated. The chemical industry remains an established political power with billions in profits at stake. Still, signs of change are everywhere.

To start with, late this spring Alar was withdrawn voluntarily by its manufacturer, Uniroyal Chemical, when the headlines and consumer complaints would not abate. Within the last couple of years, two of the industry's biggest companies, Monsanto and Du Pont, have begun to invest large amounts of research money into genetic engineering. Instead of being treated with massive doses of chemicals, some crops will now be genetically altered to increase natural resistances to pests.

By every indication, agriculture is preparing for the possibility of a postchemical era. This summer in California, the epicenter of chemical farming, the world-class Superior Farming Company harvested its first crop of organically grown table grapes. Thomas Morrison, the president of Superior Farming, says his company was responding to "consumer demand." Several other large agribusinesses are trying now to grow crops with manure and to eliminate bugs with experimental vacuum machines. Natural predators of pests, such as ladybugs and praying mantises, are also being let loose.

As for consumers, they are represented now on talk shows and in newspaper and magazine articles by the coolly passionate presence of Meryl Streep. She and Rodale's daughter Maria serve together on the board of Mothers and Others for Pesticide Limits, a group that gained prominence during the Alar scare. The April issue of *Organic Gardening* featured Streep on the cover. A meeting of two different worlds, to be sure. But it is no small matter that Streep and Rodale between them have many admirers in Congress, which next year will take up the issue of expanding the LISA program.

In all likelihood, the Bush administration will oppose such expansion. Bush's secretary of agriculture, Clayton Yeutter, placed himself squarely on the side of chemical farming in a confrontation with Hightower last January. The European Community had forced the issue by setting trade restrictions on American meat containing chemical hormones. When Yeutter announced that the United States would rather have a trade war, Hightower undercut him by promising to find enough Texans to raise chemical-free beef cattle to fill the European quota.

Bush and Yeutter were both furious with Hightower. But there is a footnote to this story that says a lot about the future of chemical farming. Chemical-free beef cattle are being raised on at least one ranch outside Santa Barbara, California. This is the Ranchman del Cielo, and its owner is Ronald Reagan.

Meanwhile, on the Flickerville Mountain Farm and Groundhog Ranch, a goodly number of pepper plants survived the slugs. As new pests make their appearance, Sinclair finds himself consulting regularly with the Rodale Institute. Every day brings another lesson in both the strengths and weaknesses of the natural order of growing things. The summer of 1989 has been a good one for Sinclair, the best in a long time.

In May there was another resignation from the *Post*. This time it was Cass Peterson, the heralded environmental writer, who, like Sinclair, was leaving at the peak of her career. Peterson is Sinclair's longtime girlfriend and partner, and she had decided to join him full-time on the farm.

An Interview with David Brower

Bill McKibben

A pair of massive incense cedars frame the view of Yosemite Falls; thousands of feet up behind them, that cascade pours over the valley wall, its misty stream blown by a cool December wind. Down on the valley floor, that breeze turns David Brower's breath white as he sits and talks in the late-afternoon sun. It is the right spot to be talking with the man who has been America's most militant and effective environmentalist in the years since World War II. The right spot because, for one thing, the American conservation movement really began here in the Yosemite Valley around the turn of the century. John Muir and his infant Sierra Club made their successful fight for a big national park—and then they waged their bitter losing battle to save the neighboring Hetch Hetchy Valley, drowned by San Francisco in the 1920s for drinking water.

Brower is the direct heir to Muir's movement. Under his directorship the Sierra Club turned from a regional into a national force in the 1950s and 1960s. He led the fight against dams on the Colorado River—a guerrilla media battle that reached its peak when one official said flooding the Grand Canyon would make it easier for tourists to visit. "Would you flood the Sistine Chapel so people could get a closer look?" the preservationists responded in full-page newspaper ads. The tactics saved the Grand Canyon but not Glen Canyon, just upstream, which was flooded to create Lake Powell and to generate hydroelectricity. "Give me back Hetch Hetchy and Glen Canyon," Brower says. "Those would satisfy me, and I'd leave the earth quietly."

If that were to happen, it would be the first time he'd ever left anything without commotion. In 1969, two years before John McPhee would enshrine him as a gruff pagan saint in his book *Encounters with the Archdruid*, Brower fell out with the board of directors of the Sierra Club. He was adamantly opposed to a nuclear reactor in California's Diablo Canyon; the board—which included his oldest friends, men like photographer Ansel Adams—voted to remove him.

He then founded a new organization, Friends of the Earth, which was designed to be international and political—two traits he thought the Sierra Club lacked. The organization quickly spread around the globe, becoming, for instance, the most powerful environmental force in Great Britain. But in this case too his brashness, and his insistence that whatever battle was at hand took precedence over fund-raising or record keeping, eventually cost him his job in the early 1980s.

He went on to found Earth Island Institute, a San Francisco–based outfit still active in environmental crusades in many countries.

Brower lives with his wife of nearly fifty years, Anne, a former book editor, in Berkeley, not far from the house where he was born or from the building where his father taught engineering at the university. At the age of seventy-seven, he continues to give speech after speech—two a week on average—all of them variations on a basic sermon he's been preaching for decades on the theme of building a sustainable society. "I'm going to Canada, the Soviet Union, Managua, Nepal and Japan in the next few months," he says [in 1990]. And he has finally finished the first volume of his massive autobiography, *For Earth's Sake*, which was published by Gibbs Smith on Earth Day.

But there's another reason that Yosemite is the perfect place to talk with Brower. In his day—the late twenties and thirties—he was one of the country's foremost rock climbers. He found nineteen new routes on the sheer granite walls of Yosemite, including the Lost Arrow Spire, a nasty steeple looming above the balcony where we are talking. He and his friends climbed New Mexico's Shiprock (you've seen it in car ads) before anyone else. He was the first to ascend many of the peaks in the Sierras. He is not as quick uphill as he once was, but on level ground he walks faster than I can.

"I get some fun at the age of 77.4 walking fairly rapidly, trying to look spry," he says. "Though I would hate to have someone apply that adjective to me. Spry? Me?"

He's willing to talk about his past—about his climbs, about the birth of the environmental movement, about battles won and lost—but he's most eager to discuss what lies ahead. "We've got to reassemble," he says. "Heal. Cure. Regenerate. You put a *re-* in front of it, and I'm for it. It's so exciting. I know I'm seventy-seven, but I've applied for a twenty-year extension."

Do you have a proprietary feeling for Yosemite, having been the first to climb so many of its walls?

I feel a little proprietary interest, as if I'd built it. It is very rewarding to come back here and say hello to the various cliffs I've known for so long and see how they're doing—and by and large they hold up pretty well.

What did it feel like to be the first man up so many mountains?

In the Sierras, I guess, I made seventy first ascents. At least I think I was the first. If the Indians did it ahead, they didn't leave any sign. Some of those were difficult. Some were quite easy, and it's just that no one had taken the trouble to go up them. My friends and I took the trouble, and it wasn't much trouble.

On the other hand, it's dirty pool to fill in the blank spaces on the map, and this was one way of filling them in. It's awfully hard to find a blank peak. Nowadays people are looking for very difficult routes up very difficult faces. But that other opportunity, to go someplace first—I've deprived them of it. If I'd done it and kept quiet about it, they could all have had the thrill of doing it the first time.

Why climb mountains? Because they're there?

I'll just reverse it—why wouldn't you? If you take any child, the instinct's there. If there's something to head for the top of, they head for it. It's the way moths head for light. I guess that it's just the moth in me. As we know now, when you get to the higher spot, you have a much better view. When you're down low, you're in most of the constituents of smog.

What element did danger play in all of this? Do people need a certain amount of danger?

I think so. I think there needs to be a certain element of risk, and that's been possibly my trouble in the environmental movement. Because I was a climber, and because there's a certain amount of risk in getting off level ground. When you get up to where the holds are very thin and there's a great deal of air under your feet, you're aware that it's not a good time to slip and that you might not survive it. . . . That just moves me to one of my favorite quotations—I'm not sure who said it: "A ship in harbor is safe, but that's not what ships were built for." I don't think we were built to be safe. I think we were built to try things.

You said that may have caused you trouble in the environmental movement?

I was perfectly willing to take some risks, and I still think it's important to be willing to do that—to make yourself a little unpopular by trying to protect something that people want to unprotect. That's not very well stated, but that's part of the reason to be in the environmental movement. You have to save things that people want to spoil. Or think that it's important for jobs or progress or whatever they think. They want to get rid of a little more wildness on Earth, and I'm in favor of saving whatever's left. We've messed up enough already. We need to save some, and I'm willing to become unpopular to do so. I wasn't too popular when

238

I lost the Sierra Club its tax-exempt status by lobbying for the Grand Canyon—but it was the right thing to do.

Do you save it for people or for its own sake?

That's the change that's gone through my mind—initially we were trying to save it for the people who want to experience it. Now I'm very much concerned with saving it for the information it contains about the way the world works. Nature is the ultimate encyclopedia. It's in many languages. Some of the languages we haven't learned to read yet. The challenge is to find out how the world works—not only up in the atmosphere, but underneath.

We have lots of species on Earth—we don't know how many. The estimates I've seen range from 5 million to 80 million. If you have a range that big, and you've only identified one and a half million, there's an awful lot we don't know. I would like to keep the opportunity to know an awful lot rather than foreclose it, which is what happens every time we move into something wild before we take inventory. If we took inventory, we would be much more careful. All I'm doing here is saying differently and at greater length what [the late conservationist and writer] Nancy Newhall said: "The wilderness holds answers to more questions than we've yet learned how to ask."

There's part of every living person that is three and a half billion years old. That's when life began, [and it] diversified immediately. A lot of structures developed; most fell by the wayside. But we are this end of that perfect transmission of information. It's a miracle. We don't need to be proud of it, because everything that's alive along with us—every tree that we're looking at, every pine, every aspen—can make the same claim. There's a little in it that's been passed along for three and a half billion years. And I don't just like to mess, any more than we already have, with what's made this possible. Because I think it's produced some pretty nice things, including us.

At what point did you begin to evolve into an environmentalist?

The first thing that bothered me was when I was quite young. Our family would take camping trips into the Sierra. We'd go over what was then the Lincoln Highway—later Highway 40 and I-80. But when it was just this simple, one-lane dirt road, we passed through some very nice forests. I remember this one particular spot where I went into this forest and discovered a spring.

I came back a few years later, and the place had been logged. There was no forest to speak of, and there was no spring. And I felt deprived and angered—who the hell did this to my forest? That was the first time I was annoyed. I was eight or nine. And there were plenty of things to annoy me as I got older. And there are plenty of things now—there is no shortage of things that annoy me about what people want to do with a beautiful piece of earth. Just about anything a developer wants to do can annoy me. And the addiction to automobiles that no one wants to tackle. I suppose the idea of growth is somehow underneath all of this. Somehow we have this strange idea that everything needs to grow, and it doesn't. Children need to grow, and old people need to be edited out—and that's all arranged for us.

But anyway, my chagrin at the destruction of that little piece of forest was the first time. It didn't move me into doing anything about it until I had joined the Sierra Club and was just going out on the trails. In 1935, when I came to Yosemite Valley, I came up with the bright idea that there should be a cable car from Mirror Lake to Mount Hoffman, so that we could get to some real ski country. I got scolded for thinking that, and the scolding took. There are plenty of things you can do mechanically, but there are plenty of places that need to be unspoiled. And I decided then and there that I was in favor of unspoiling places.

When you give your basic sermon, what are the parts that seem to really grab people?

The part that grabs people the most is [when] I squeeze time. I squeeze the age of the Earth, four and a half billion years, into the six days of creation. The Earth is created Sunday midnight. Life appears Tuesday noon. Saturday morning, early, there's been enough chlorophyll at work that fossil fuels can begin to form. At four in the afternoon the great reptiles come onstage; at nine in the evening they're finished. Eighteen minutes before midnight the Colorado River starts digging the Grand Canyon.

Four minutes before midnight something like us—though not really like us—starts to appear. One and a half seconds before midnight we invent agriculture. In the next half second we're so successful that the forests ringing the Mediterranean have disappeared, and with it the Sumerian civilization. As it dies, a bristlecone pine sprouts on White Mountain in California, and it's still alive. A third of a second before midnight, Buddha. A third of a second, Christ. A fortieth of a second before midnight, the Industrial Revolution. An eightieth of a second, we discover oil. A hundredth of a second, Ross and Mary Grace Brower discover David Brower. In less than his lifetime, the population of the Earth trebles, and the appetite for resources quadruples. Since then we've used up more resources than were used up in all previous history. And now it's midnight, and most world leaders think that this can go on at an accelerating rate—or, to say it another way, that we can multiply and subdue the Earth but do it even faster. This is a bad idea. We've got to come up with some good ones.

This hundredth of a second—is it a good time to be alive?

I myself would like to go back to the time of Moses and say, "Go back up and bring back down the other tablet!" The Ten Commandments just talk about how we're supposed to treat each other. There's not a bloody word about how we're supposed to treat the Earth. And we won't have each other without an Earth. It must be up there still. Find that other tablet.

Do you have religious instincts?

I'm a dropout Presbyterian. I've come to the conclusion that the religions, every one of them, have been devised by us to help us rationalize how things work or things we can't understand. Where we say we are made in God's image, it's just the same as saying God's made in our image, which is part of our hubris. I just settle on admiring the system, however it came about. That's what I admire, that's what I trust. That's what I don't want to mess up. I guess I'm just a naturalist.

You've been a naturalist longer than just about anyone else around. How has the environmental movement changed over the decades?

The main thing that's happened is that the membership has risen dramatically. When I joined the Sierra Club, there were 2,000 members. Now there are 550,000. At that growth rate we'll soon have far more members than there are people on Earth. The other thing that's happened is the rise of environmental litigation—that's opened the door for a lot of wonderful victories. What's got to come now is environmental politics; we've got to green the politics of this country so environmentalists can get into office.

You lost Glen Canyon to a dam because the Sierra Club was willing to compromise to prevent other dams. What did it teach you about compromise as a tactic?

I think compromise is a necessary part of society. We hire people to compromise, and we send them to Congress. But an organization like the Sierra Club or Friends of the Earth or Earth Island Institute—that's not our job. Our job is to find the strongest arguments we can for leaving things as they are, to make the argument that all the compromises have been made already. We've made too many, and the Earth is suffering. And we're going to stand fast. By standing fast, at least we'll give the people who are paid to do the negotiating some grounds to do the negotiating from. If we compromise, then the final result is between our compromise and total devastation, instead of between total protection and total devastation. They always want just one more dam. And the result is that, of the 6,000 miles of salmon streams in the Imperial Valley when I arrived in this world, there are only 300 left.

You've had a very stormy relationship with environmental groups, considering you're one of the foremost environmentalists of your time. What happens?

[Supreme Court justice William] Douglas put it very well to FDR: Every government bureau that reaches the age of ten should be abolished. After that, it's more concerned with itself than with its mission. And I think that becomes true of people and organizations beyond government bureaus. You get comfortable where you are, you get practical. I like the definition of the practical man: A practical man is one who has made all his decisions, has lost the ability to listen and is concerned to perpetuate the errors of his ancestors. And that begins to build itself into us. But somehow I missed that virus—I've got lots of others, but I missed that one. I was taught a lesson that if you hang in there tough, you can save things.

The kind of movement that saved the Grand Canyon—is it capable of addressing global problems like carbon-dioxide emissions?

I think it is. The whole thing—the excitement over global warming, acid rain, hole in the ozone, loss of species—is getting through the way it never did before. People are always looking for excuses, for ways to deny things. But it's getting harder.

I think one of the things that is getting through very slowly is the problem of global inequality. We've been so enraptured with getting a high standard of living that we haven't bothered to find out what it cost the rest of the world. We in the West have ransacked four-fifths of the world. That's a little harder to break away from. We're so dependent on it. We need all these slaves.

Many would say it's environmentalists who are dooming the third world to poverty forever because they want to slow economic growth.

That's balderdash, of course. The people who say that are people who, with their advice and practices, have been widening the gap between haves and have-nots. I've been watching that widen for the last twenty years. I remember people saying we must have economic growth. And we've had economic growth, and the gap is wider. We have more homeless than we ever had. The third world is in trouble in no small part because we make demands on their resource base for products that bring them cash, which they can't eat.

Environmentalists are also often accused of being elitists, of wanting to set aside beautiful places for themselves.

Those statements are made by people who are high up in the *Fortune* 500. They call us *elitists*—them with their clubs and two summer houses. I like [environmental writer] Garrett Hardin's line: Don't call me names; tell me where I'm wrong. What we're locking out [of wilderness areas] are the destructive forces—the chain saws, the bulldozers, the pourers of concrete, the poisoners, the compactors. That's what we're locking out. We're not locking people out. There's some people who can't get there, if they're old or infirm. But I certainly, at the age of 77.4, don't want people to make it any easier for me to get into any wilderness area anywhere. Because I cannot, in any conscience, take away the chance for somebody to earn it instead of having it spooned out. We're supposed to leave some choices for the next generation—and we're leaving fewer and fewer choices.

Who are the villains? Is it just greedy corporations?

I think there is corporate greed, and I think the corporate structure, as it now exists, is a device for separating people from their consciences and protects them when they do things their conscience tells them are wrong. They're still safe in court. I'd like to see that barrier between corporation and conscience removed. I have nothing against making a profit—you need an accumulation of something to work with to do something else you need to do. I'm not going to attack the almighty dollar. If making some money is going to encourage people to do things, fine, but I want it to encourage them to do good things. That's what concerns me—the impact of their avariciousness.

But aren't we biologically programmed, like lemmings, to keep increasing our numbers until we run out of resources?

We're not programmed to do that. Somewhere along the line we picked up the ability to cut down the death rate. We devised ways of reducing our death rate severely, and that has overloaded the Earth with people. But there is nothing I can think of to do about it, except to reason our way out of having as many people as we can now have. With the death rate controlled, we have to control the birth rate. But that is a hard thing to get people to understand, including the pope.

Do you think there's a chance we'll wise up as a species?

I think there is a chance. We've been going through a long period where we've celebrated our aggressiveness, but we also have other capabilities. We are able to live in symbiosis, to live in peace. The biggest problem with peace, I think, is that it tends to be a little dull. We have to find other ways to get our excitement rather than killing people. There must be some other sport, some other form of keeping interest up. It sounds silly, but if you go through the history books, how many pages are devoted to the wars and how many to the peaces? We do have compassion, we do care about other people. Get people under an awning in a storm, and they're friendly. There's something else outside themselves to be concerned about, and they don't mind each other for that brief moment. We have to do this globally somehow. I hope we'll get a fuller public understanding of the storm that is besetting the Earth; perhaps that will do the trick.

Is it starting to happen?

I think so. We usually wait to the last minute. As a whole, civilization is as good at procrastinating as I am. And now, as we face problems like global warming, we're about at the last moment, and we better start to get along. I'll go back to Pogo's argument: We've met the enemy, and now we know who it is. I like the other Pogo line: We are confronted with insurmountable opportunity. As a mountaineer, I've declared many things insurmountable, and they've been climbed. I've climbed some of them. That's the thing to do, just figure out how to get there. That can be part of the excitement. That's my new rage—how to put things back together. There's a man in Costa Rica who is restoring a dry rain forest— that's the most exciting thing I can think of. And we've got $150 billion worth of work to do trying to restore the damage we've done with nuclear waste. That should keep us busy a little while.

What about George Bush? Is he getting the message?

I think his attitude is one not even of benign neglect—it's just neglect, period. Rip Van Winkle has nothing on George Bush. I don't think there's any sign of him becoming the environmental president he says he is. No sign at all—just the lips. I hope his wife can shake him up—maybe *he* needs an astrologer.

You've been described in the past as America's most radical environmentalist.

The people who are radical are the people who are trying to destroy the life-support system. They're the ones who are antipeople, anti-Earth, antinature, antisurvival. These people are not called names, except by a few of us. But these are the people you've got to watch out for. These people who call those who want to save nature radical are making groups like Earth First! necessary. If they don't like Earth First!, they should stop doing what they are doing to make it necessary.

What do you think about militant groups like Earth First! that spike trees to keep them from being cut down and destroy property in order to save wilderness?

I think Earth First! is right on target. I just wish they'd use biodegradable spikes in trees. The destruction going on in forestry practice is a major deprivation of the life-support system right now. It is terrorism. It is ecological terrorism—what is being done to the life-support system, the life forms there we don't understand. If you settle on a nice round number of 50 million species and divide that by the rate with which we are getting rid of species and realize that that number is closely related to getting rid of tropical rain forests, where half the species are presumed to be, you come to the answer that we're losing one species a minute. If 50 million species is too high and you say 25 million, you're losing one every two minutes. If it's only 5 million, one every ten minutes—if that makes you feel any better. It doesn't make me feel any better.

What are we doing to the system? We're taking parts out—maybe we're taking the carburetor out. Then what are we going to do? If we're just taking the spokes out of a wheel, we can get away with it for a while. But if we take out the tire, we're in deep trouble. It might be the thing that will make all the dominoes fall. So it's stupidity we're practicing and admiring, while calling radical the people who oppose it.

One can't advocate violence against property—it's against the law. But I remember [essayist and novelist] Ed Abbey's advice: Follow your conscience.

If you had to give your vision, what would a sustainable society look like?

We'd reduce our population. If we can't support 5 billion without mining, without irreparably damaging, the life-support system, then we'd better start moving by some less precipitate method than having mass starvation. Go for a lower number, and concurrently cut the appetite for destroying resources. That can be done. We can live more lightly on Earth and leave it in better shape than we do. . . . And then we've got to use our technology to put things back together. There are some things you can't put back together. If you destroy a species, that's it. We've destroyed millions—that's it. *We* did it—the evolutionary force didn't do it.

Why don't we just invent some new species?

Genetic engineers frighten me just as much as nuclear engineers. We've got to put limits on science. You don't put together something you can't take apart, and you don't take apart something you can't put together—though it offends most scientists right now to think about that.

It's not that I abhor science. We're going to need science to put things back together. We're going to need chemists to concoct the antidotes to what chemists have given us already. I can't do it. We've got to take all this knowledge and keep the Earth together with it instead of disassembling it. Let's reassemble, everyone. That's challenge enough. If we start to do that, we'll find it's necessary to recycle everything, as nature does. And we can do that. Let the market system meet that challenge—to serve a sustainable society and not a disappearing society. Rome fell, the Sumerian civilization fell—let's see if *we* can keep upright.

But what's wrong with getting rid of something small like the snail darter [a small endangered fish whose existence almost blocked the construction of a major dam in Tennessee], which doesn't seem to have much place in the scheme of things?

I have no idea. I don't know what the snail darter's function is.

What if I proved to you that it would have no effect?

I'll take the snail darter's vote first. What does it think of being gotten rid of?

So things have some sort of intrinsic value?

I think so. I'm a very poor disciplinarian—I think our dog has rights. I refuse to give our dog orders I think are demeaning. It was very easy when we had a monkey—there was no infringing on Isabelle's rights in any way.

Most of us who believe this stuff are also hypocrites of a large order. We drive around in cars, use more power than we probably should. How do you deal with that, with the hypocrisy of your own life?

I don't know. I confess who the worst hypocrite is I know—I look in the mirror each morning and see him. I can list the things I have, some of which aren't necessary, some of which I think are necessary, some of which I fool myself into thinking are necessary. We have lots of company in our hypocrisy. We should try and get a lot of company in going the other way. I will stop driving my car if you'll stop driving your car. If you don't want a Mercedes, I won't get one. We need a little more company in demanding less of the Earth—to make it the popular thing.

I do think it's important for the people who want to lead not to appear too oddball about it. Now, I'm an oddball telling people not to be oddballs. If you're going to be an oddball, wear a three-piece suit and a tie now and then! Don't look too far-out so you scare people! It's not too big a sacrifice for the people who want to reform to give up looking scruffy or tatty or whatever it is.

What is it that's special about human beings as a species?

I'll just repeat the remark of my son, Kenneth: If we hadn't made the mistake of trading in flippers for thumbs, we might have gone on for another 30 million years without blowing the place up. Where did we go wrong? That's where we went wrong [*laughs*].

We are, I think, quite special. We do have this grasp of what's going on in the universe. I like the greater grasp that comes from the Eastern religions. I find that an old joke of mine is consonant with their idea of the universe: God heard the scientists talking about the big bang. And God said, "Which big bang?"

We've been around for about 300,000 years as *Homo sapiens*. And that's nothing. And we're about to wipe ourselves out, because we're so bright we're going to drown ourselves in our own cleverness. We can do it. We're well on our way.

And can that cleverness be our way out as well?

That's my hope, my belief—the only reason I can be an optimist at all. I'll take that choice, because the other choices are not acceptable. So let's go for it. Try to put out the challenge—who's got the best plan for a sustainable society? And I don't mean sustainable for a week. Let's see if we can stick around for as long as the dinosaurs did. We've got a long way to go.

Heaven Is Under Our Feet: Voices for Walden Woods

*E*ver since 1845, when Henry David Thoreau began to build his cabin by the side of Walden Pond, the surrounding woods have been hallowed ground for the environmental movement and American literature. Although much of Walden Woods has survived relatively intact since Thoreau's day, two sites—Brister's Hill and Bear Garden Hill—are currently threatened by real-estate developments that would irreparably damage the beauty and ecological integrity of the woods. In 1990, Don Henley founded the Walden Woods Project to raise funds to purchase and protect the two parcels of land. As part of that effort, the project is publishing *Heaven Is Under Our Feet: A Book for Walden Woods*, a collection of essays by environmentalists, writers and concerned figures in the worlds of politics and entertainment. All royalties and a portion of the proceeds from the book will go to the Walden Woods Project.

We've chosen three essays from *Heaven Is Under Our Feet*: Henley's essay serves as an introduction and statement of purpose; the novelist E. L. Doctorow's essay, taken from his remarks at the press conference announcing the Walden Woods Project, speaks of how Thoreau's writings give Walden Woods a "historical luminosity"; and the poet and essayist Robert Bly praises Thoreau's uniquely intimate connection to the land—ED.

Don Henley

I GREW UP outside. Outdoors and outside. I liked my home and my little town, but in the company of other humans, I often felt like an alien—as if I had been dropped in that place by space travelers that I couldn't remember. It wasn't a bad place, really. It was quite beautiful in the spring and fall but pretty bleak in winter because it rarely snowed, which left the frozen earth in a dreary state of gray-brown nakedness. Still, I found comfort and wonder outdoors in all seasons. In spring I would don my raincoat and go walking in thundershowers. This little custom scared the living daylights out of my mother, who was certain I would be struck by lightning or swept away by the malignant elements. But I loved weather. Still do. Storms heighten the senses. You know you're alive.

In summer I roamed the woods with my dog. There were streams and ponds, tadpoles and frogs and chameleons to catch—typical Mark Twain, Norman Rockwell boyhood stuff. When I got a little older and learned how to use a rod and reel, Dad began to take me on fishing trips to Caddo Lake, a large, elongated body of water that lies half in Texas and half in Louisiana. It is one of those atmospheric Southern places populated by cypress trees dripping with Spanish moss. It is also home to a good many alligators, pelicans, catfish, perch and bass. This was my Walden. I caught my first fish there.

Along with his love of lake fishing, my father was an avid gardener. On many a summer morning he rousted me out of bed well before sunup and handed me a hoe. We had more than an acre to tend, and the objective was to get as much as possible done before the sun got too high in the sky and the temperature rose above 100. The humidity in that region, while good for the skin and for growing vegetables, is oppressive, and heat exhaustion is always a possibility in summer. On several occasions my thoughts turned patricidal. But as the years have passed, I have grown to appreciate what my dad taught me, not only about growing things in the earth but also about responsibility and the value of hard, physical work. I now derive physical and spiritual pleasure from gardening. All this galls me a little, because my dad always said it would turn out this way.

I honestly don't remember when I was first introduced to the works of Henry David Thoreau or by whom. It may have been my venerable high-school English teacher, Margaret Lovelace, or it may have been one of my university professors. Thoreau's writing struck me like a thunderbolt. Like all great literature, it articulated something that I knew intuitively but could not quite bring into focus for myself. I loved Emerson, too, and his essay "Self-Reliance" was instrumental in giving me the courage to become a songwriter. The works of both men were part of a spiritual awakening in which I rediscovered my hometown and the beauty of the surrounding landscape and, through that, some evidence of a "higher power," or God, if you like. This epiphany brought great comfort and relief, because the Southern Baptist Church just wasn't working for me.

I have volunteered all this because, as Thoreau declares in the beginning of *Walden*, "I should not talk so much about myself if there were anybody else whom I knew as well. . . . Moreover, I, on my side, require of every writer, first or last, a simple and sincere account of his own life, and not merely what he has heard of other men's lives." Also, there has been a great deal of curiosity, speculation and, in some quarters, skepticism bordering on cynicism as to how and why I came to be involved in the movement to preserve the stomping grounds of Henry David Thoreau and his friend and mentor Ralph Waldo Emerson. What, in other words, is California rock and roll trash doing meddling in something as seemingly esoteric and high-minded as literature, philosophy and history— the American transcendentalist movement and all its ascetic practitioners? Seems perfectly natural to me. American literature, like the air we breathe, belongs—or should belong—to everybody. I'm an Everyman kind of guy. I have had a respectable degree of success communicating with him for the past twenty years. In short, there is a job that needs doing, needs some "plain speaking," and I think I can help—even from here in Gomorrah-by-the-Sea. Indeed, living and working in Los Angeles has taught me a great deal about the stormy confluence of art and commerce—about how the "real world" operates. And the preservation of historic Walden Woods is going to require a healthy dose of operating in the real world. The great halls of learning may keep Thoreau's literature and principles alive, but they will be of little help in fortifying the well whence they sprang.

Unfortunately, the focus of preservation efforts has come to rest on Walden Pond and its immediate surroundings. That is all well and good, except that there remain approximately 2,600 acres that are inside the historic boundaries of Walden Woods and deserve protection as well. Thoreau did not live *in* Walden Pond; he lived beside it. The man did not walk on water; he walked several miles a day through the woods, and his musings and writings therein figure at least as prominently in his literature as Walden Pond does. Walden Woods is not a pristine, grand tract of wilderness, but it is still, for the most part, exceedingly beautiful and inspiring. It is, for all intents and purposes, the cradle of the American environmental movement and should be preserved for its intrinsic symbolic value or, as Ed Schofield, Thoreau Society president, so succinctly puts it: "When Walden goes, all the issues radiating out from Walden go, too. If the prime place can be disposed of, how much easier to dispose of the issues it represents."

People seem to be very attached to their symbols, although sometimes they cannot articulate why. (Witness the brouhaha over flag burning.) In the United States we revere the flag, the cross, the Star of David, the Mississippi River, the Grand Canyon. These things tell us who we are as people; they tell us where we stand in the world, in the universe; and they show us that there is ultimately something larger and more important than ourselves. Being the proud nation that we are, that's a hard one for us to swallow. Symbols, therefore, get perverted and their meanings twisted because we fail to see the connectedness of things. We, the thinking, reasoning animal—the "highest" form of life on the planet—tend to view ourselves as outside of or *above* the natural order. Historically, we have often been diametrically opposed to it. Wallace Stegner, in his book *The American West as Living Space*, writes:

> Behind the pragmatic, manifest-destinarian purpose of pushing western settlement was another motive: the hard determination to dominate nature that historian Lynn White, in the essay "Historical Roots of Our Ecologic Crisis," identified as part of our Judeo-Christian heritage. Nobody implemented that impulse more uncomplicatedly than the Mormons, a chosen people who believed the Lord when He told them to make the desert bloom as the rose. Nobody expressed it more bluntly than a Mormon hierarch, John Widtsoe, in the middle of the irrigation campaigns: "The destiny of man is to possess the whole earth; the destiny of the earth is to be subject to man. There can be no full conquest of the earth, and no real satisfaction to humanity, if large portions of the earth remain beyond his highest control." . . . That doctrine offends me to the bottom of my not-very-Christian soul.
>
> Our very virtues as a pioneering people, the very genius of our industrial civilization, drove us to act as we did. God and Manifest Destiny spoke with one voice urging us to "conquer" or "win" the West; and there was no voice of comparable authority to remind us of Mary Austin's quiet but profound truth, that the manner of the country makes the usage of life there, and that the land will not be lived in except in its own fashion.

In the Book of Genesis, God said, "Let [man] have dominion over the fish of the sea, and over the fowl of the air, and over the cattle, and over all the Earth, and over every creeping thing that creepeth upon the Earth." If the Lord did in fact utter these words sans any kind of cautionary caveat, then they must have lost something in the translation, because we have just about "dominioned" ourselves to death.

God, the Divinity, the Oversoul (or whatever term you wish to use) is manifest nowhere as surely and as magnificently as in untrammeled nature, and I think that a great deal of the spiritual groping and confusion in our society stems from the fact that we have strayed so far from our roots in the land. We have distanced ourselves from contact that we once had on a regular basis with the natural cycle of birth, death, decay and rebirth.

I'm not saying that everybody should pack up and move back to the country. The country couldn't take the onslaught. What I am saying is that we should make an effort to rekindle respect for the values that come from life lived in harmony with the land.

What we lack is humility, and underlying that is a dearth of self-esteem. From the endless stream of macabre, self-congratulatory pep rallies for returning soldiers, to the heedless destruction of our ancient redwoods (the oldest living things on the planet), to the building of more powerful weapons of war, to the fevered paving over of precious farm-land, the construction of impersonal malls and ever-higher skyscrapers and the slaughter of defenseless animals for the manufacture of vanity items such as makeup and fur coats, it appears that we, as a people, must go to greater and greater lengths to feel good about ourselves. "What we call man's power over nature," writes C. S. Lewis, "turns out to be a power exercised by some men over other men with nature as its instrument." Calvin Coolidge once said that the business of America is business. This, I fear, will finally be our undoing unless we change our current direction and thought. In the ongoing battle between commerce and the natural environment, nature is almost invariably the loser. As Stegner writes:

> Habits persist. The hard, aggressive, single-minded energy that according to politicians made America great is demonstrated every day in resource raids and leveraged takeovers by entrepreneurs; and along with that competitive individualism and ruthlessness goes a rejection of any controlling past or tradition. What matters is here, now, the seizable opportunity. "We don't need any history," said

one Silicon Valley executive when the Santa Clara County Historical Society tried to bring the electronics industry together with the
few remaining farmers to discuss what was happening to the valley
that only a decade or two ago was the fruit bowl of the world.
"What we need is more attention to our computers and the moves
of the competition."

A high degree of mobility, a degree of ruthlessness, a large
component of both self-sufficiency and self-righteousness mark the
historical pioneer, the lone-riding folk hero, and the modern busi-
nessman intent on opening new industrial frontiers and getting his
own in the process.

Stegner goes on to lament the disappearance of the true culture
hero—"the individual who transcends his culture without abandoning
it, who leaves for a while in search of opportunity but never forgets where
he left his heart."

Yet another problem, and perhaps the most insidious of all, is *denial*.
One hears that word quite often of late. In my vicinity it is generally
used in connection with alcoholism and drug abuse. It is now, however,
more than appropriate to describe the state of American or, in most cases,
global consciousness concerning the crisis facing the ecosystem. Where
the environment is concerned, according to reams of scientific data (any
sane person can see it with his own eyes), there is definitely an elephant in
the room. Unfortunately, there are still an alarming number of people,
including a great many officials in our federal government, who insist
there isn't. "Further studies," I believe, is the popular talismanic phrase.

In *The True and Only Heaven*, Christopher Lasch writes that we seem
incapable of dealing with the hard choices posed by environmental
issues. "Both the right and left," he writes, "therefore prefer to talk
about something else—for example, to exchange accusations of fascism
and socialism. But the ritual deployment and rhetorical inflation of these
familiar slogans provide further evidence of the emptiness of recent
political debate."

This book, then, is written by those who realize that ultimately our
best hope for salvation is ourselves and that we must communicate with
one another in whatever forum is open to us—or as James Hillman put

it, we are born, first and foremost, citizens, and we abandon citizenship when we stop talking. Guided by the wisdom of history, we will, I hope, be able to regain a sense of our inextricable linkage to the natural world and a recognition of the value of the symbols and signposts that point the way.

E. L. Doctorow

THOREAU'S WALDEN; OR, Life in the Woods, like Twain's Huckleberry Finn or Melville's Moby Dick, is a book that could have been written only by an American. You can't imagine this odd, visionary but very tough work coming out of Europe. It is peculiarly of us; it is indelibly made from our woods and water and New World ethos. But more than that, it is one of the handful of works that make us who we are. Walden is crucial to the identity of Americans who have never read it and have barely heard of Thoreau. Its profound complaint endures to our century—for example, it was the text of choice in the 1960s, when the desire to own nothing and live poor swept through an entire generation.

It is a sometimes prickly book about independence and a practical how-to book on the way to live close to the earth in a self-sufficient manner; it is a sometimes philosophical book about values—what we need to live in self-realization and what we don't need, what is true and important, what is false and disabling—and it is a religious book about being truly awake and alive in freedom in the natural world and living in a powerful transcendent state of reverence toward it. Presented as the story of Thoreau's life at the pond over a period of two years, Walden fuses his political, economic, social and spiritual ideas in a vision of supreme common sense.

All right, that begins to describe the book. What about the place? We have the book—why do we need the place?

Literature, like history, endows places with meaning, and in that sense it composes places, locates them in the moral universe, gives them a charged name. So in effect literature connects the visible and the invisible. It finds the meaning, or the hidden life, in the observable life.

It discovers the significant secrets of places and things. That is what makes it so necessary to us; that is why we practice it. Uncharged with invisible meaning, the visible is nothing, mere clay, and without a visible circumstance, a territory, to connect to, our spirit is shapeless, nameless and undefined.

"Near the end of March, 1845," Thoreau writes, "I borrowed an axe and went down to the woods by Walden Pond nearest to where I intended to build my house, and began to cut down some tall arrowy white pines." Walden is the material out of which Thoreau made his book—as surely as he made his house from the trees he cut there, he made his book from the life he lived there. The pond and woods are the visible, actual, real source of his discovered, invisible truths, the material from which he made not only his house but his revelation.

That Walden is a humble place—an ordinary pond, a plain New England wood—is exactly the point. Thoreau made himself an Everyman and chose Walden for his Everywhere.

Clearly, there is a historical luminosity to these woods. They stand transformed by Thoreau's attention into a kind of chapel in which this stubborn Yankee holy man came to his and, as it turns out, our redemptive vision. So there is a crucial connection of American clay and spirit here: If we neglect or deface or degrade Walden, the place, we sever a connection to ourselves, we tear it asunder. Destroy the place and we defame the author, mock his vision and therefore tear up by the root the spiritual secret he found for us.

We need both Waldens, the book and the place. We're not all spirit any more than we are all clay; we are both and so we need both—as in: You've read the book, now see the place.

You have to be able to take the children there, and to say: "This is it, this is the wood Henry wrote about. You see?" You give them what is rightfully theirs, just as you give them Gettysburg because it is theirs.

But in fact you don't even have to see the place as long as you know it's there and it looks much as it looked when he was cutting the young white pines for his house. Then it is truly meaningful in spirit and in clay—like us, and like the world invisibly charged with our idea of it.

Robert Bly

THOREAU WAS CAPABLE of true patience in observing the nonhuman world, and he exclaims in one passage, "Would it not be a luxury to stand up to one's chin in some retired swamp for a whole summer's day?" If we've read Thoreau, we know that he would be perfectly capable of it. He walked two hours each day and noted with the most astonishing perseverance and tenacity the exact days on which wildflowers—dozens of varieties—opened in the forest. In 1853 he wrote in his journal: "My Aunt Maria asked me to read the life of Dr. Chalmers, which, however, I did not promise to do. Yesterday, Sunday, she was heard through the partition shouting to my Aunt Jane, who is deaf: 'Think of it: He stood half an hour today to hear the frogs croak, and he wouldn't read the life of Chalmers.' "

Thoreau felt invited to observe the detail in nature, and he did not receive this invitation from Wordsworth or Milton; it came to him as a part of his genius. When he was twenty-one, he wrote in his journal: "Nature will bear the closest inspection. She invites us to lay our eye level with her smallest leaf, and take an insect view of its plain." There is something brilliant in the last clause, advising us to take a low-lying or insect position when we look. One day, while he lay on his back during a soaking rain, he saw a raindrop descend along a stalk of the previous year's oats. "While these clouds and this sombre drizzling weather shut all in, we two draw nearer and know one another," he wrote in his journal. R. H. Blyth declared this sentence to be one of the few sentences in the English language that are genuine haiku.

We need to understand that Thoreau received through Emerson and Coleridge, through the Eastern spiritual books he read, among them those of the Indian poet Kabir, and through Goethe, Schelling and other German writers, the doctrine that the spiritual world lies hidden in—or moving among, or shining through—the physical world. Nature is one of the languages that God speaks.

Since the physical world conceals or embodies a spiritual world, if one studies facts in nature, one might be able to deduce or distill from many

258

physical facts a spiritual fact. Thoreau remarks in his journal on February 18, 1852, "I have a common-place book for facts, and another for poetry, but I find it difficult always to preserve the vague distinction which I had in mind, for the most interesting and beautiful facts are so much the more poetry and that is their success. They are translated from earth to heaven. I see that if my facts were sufficiently vital and significant— perhaps transmuted into the substance of the human mind—I should need but one book of poetry to contain them all." So one can translate certain facts "from earth to heaven." Scientists, because they do not know Kabir's truth of the double world, do not translate. Scientific study of facts in Thoreau's time did not encourage the scientist to cross over the threshold between worlds. But Thoreau is able to cross from earth to heaven: "I see that if my facts were sufficiently vital and significant— perhaps translated into the substance of the human mind"—they would become poetry.

We understand that Thoreau's observation is not a simple-minded cataloging of detail. Behind his persistence lies the promise, grounded in his vast reading, that, in Coleridge's words, "each object rightly seen unlocks a new faculty of the Soul." What is it like, then, to look at an object rightly? Suppose one watched ants fighting. Eyes see surprises, polarities, nuances; the observer's language, if he or she wrote of the battle, would have to contain those nuances, so that the reader could also see rightly. We notice in the following passage that Thoreau provides "embraces," "sunny valley" and "chips" as nuances among the violence: "I watched a couple that were fast locked in each other's embraces, in a little sunny valley amid the chips, now at noonday prepared to fight till the sun went down, or life went out. The smaller red champion had fastened himself like a vise to his adversary's front, and through all the tumblings on that field never for an instant ceased to gnaw at one of his feelers near the root, having already caused the other to go by the board; while the stronger black one dashed him from side to side, and as I saw on looking nearer, had already divested him of several of his members. They fought with more pertinacity than bulldogs."

How good "pertinacity" is here! The swift changes of mood in animal encounters, the intricacy of instinctual gesture, the mixture of comic and tragic, require a vocabulary that can go from high to low in an instant, that can move from dark to light and back, from metallic word to fragrant word, from a slang phrase to words from the Middle Ages or the eighteenth century. American democracy suggests that good writing about nature requires only a simple heart, but bravery of soul, immense learning and cunning in language—none of them simple—are what nature writing requires.

We recognize that Thoreau's account of the ant battle is not pure observation without human imposition; while he observes details, he is also declaring that men's proclivity is mechanical and antlike: "Holding a microscope to the first mentioned red ant, I saw that, though he was assiduously gnawing at the near foreleg of his enemy, having severed his remaining feeler, his own breast was all torn away, exposing what vitals he had there to the jaws of the black warrior, whose breastplate was apparently too thick for him to pierce; and the dark carbuncles of the sufferer's eyes shone with ferocity such as war only could excite."

"Assiduously" is essential here. Long, "unnatural" words suggest the fierce intensity of the insect world. Thoreau places "feeler," a word a child might use, near "vitals," an adult word that evokes complicated feelings, including fear. He mingles with that "black warrior" and "breastplate," words that carry a Middle Ages fragrance, and they prepare for the astonishing phrase "dark carbuncles of the sufferer's eyes."

Thoreau attempts something new in American literature. He does not agree with earlier New Englanders that the world is fallen and a dark ruin but believes by contrast that the world remains radiant from the divine energy that shines through it. A few days before he died, a family friend asked him "how he stood affected toward Christ." Thoreau answered, as reported in the *Christian Examiner* in 1865, that "a snowstorm was more to him than Christ." He is suggesting, I think, that even in the 1860s, so far into the nineteenth century, the snowstorm is still luminous with spiritual energy, and Christ is not needed to lift it back up into radiance. The snowstorm and God had never quarreled.

Thoreau trained himself over many years to see. His training involved a number of disciplines. The first was constant labor. His journals are so immense that they must have required, during his short life, two or three hours of writing each day, over and above the walks he wrote about. Second, he aimed to become just, and in this struggle followed the ancient doctrine, contrary to scientific doctrine, that certain secrets of nature reveal themselves only to the observer who is morally developed. The alchemists founded their penetration of nature on their moral character. Concentrating on a "low-anchored cloud," Thoreau wrote:

> Drifting meadow of the air
> Where bloom the daisied banks and violets,
> And in whose fenny labyrinth
> The bittern booms and heron wades;
> Spirit of lakes and seas and rivers,
> Bear only perfumes and the scene
> Of healing herbs to just men's fields!

Chemical Manufacturers Association,
55
Chenegans, 153–55
Cherney, Darryl, 112, 113
Chesapeake Bay, 215
Chlorofluorocarbons, 3, 7, 33, 40, 57
Citizen's Clearinghouse for Hazardous
Waste, 222
Clark, Mike, 40
Clean Air Act, 23, 26, 29–32,
52–58, 180, 218
Cline, Dave, 138, 180
Clorox Company, 204–205
Coastal Waters in Jeopardy, 168
Coast Guard, U.S., 46, 119, 120,
122, 129–31, 135–37, 143,
159, 176–77
Cockburn, Alexander, 106
Collor, Fernando, 97
Commoner, Barry, 13
Copeland, Thomas, 149–51
Corcovado National Park, 70
Cordova District Fishermen United,
131, 159
Council on Competitiveness, 24
Cousins, Gregory, 120–21, 136
Cowper, Steve, 132, 138, 177
Crews, Harry, 127

Daime, 98–104, 105
DDT, 123, 217, 231
Deakin, Edward B., 125–26
Dean, Norman, 10
Dedman, Palikapu, 77
Deforestation, 60, 87, 89, 94,
111–13
Denison, Richard, 220
Department of Agriculture, U.S., 40,
46, 232
Department of Defense, U.S., 38, 40,
43–46, 48, 49–50
Department of Energy, U.S., 38, 40,
41–43, 48–50
Department of Interior, U.S., 135,
137, 164, 165–66

Department of Transportation, U.S.,
28
DeYoung, Karen, 4, 7
Dillman, Dick, 192–93
Dingell, John, 31–32
Doctorow, E. L., 256–57
Dolphins, 102, 199–200, 206
Dow Chemical, 55, 229, 232
Drexel Burnham Lambert, 108, 109
Du Pont, 33, 34, 55, 57, 233
Dykstra, Peter, 8–9, 179, 202, 203

Earth Day, 15–16, 18, 20, 22, 51, 55
Earth First!, 112, 204, 206, 245–46
Earth Island Institute, 199–200, 203,
236
Eckart, Dennis, 43
Egan, Bill, 164
Ehrlich, Paul, 20
Eli Lilly, 87
Emerson, Ralph Waldo, 251, 252,
258
Encounters with the Archdruid (McPhee),
236
Environmental Action, 7, 203
Environmental Defense Fund, 216–23
Environmental Protection Agency, 5,
24, 39, 46, 47–48, 50, 53–57,
137, 177, 180, 218
Erickson, Gregg, 127, 138, 181–82
Extinction of species, 86, 106, 246,
247
Exxon Valdez oil spill, 115–86, 205
cleanup, 128–35, 142, 143–45,
149–50, 152–53, 174, 183
fishermen and, 158–66
the future, 174–86
recounting of, 119–22, 128
wildlife affected by, 145–49

Farming, 227–34
Fate of the Forest (Hecht and Cockburn),
106
Federal Aviation Administration, 46

Balancing on the Brink of Extinction: The Endangered Species Act and Lessons for the Future
Edited by Kathryn A. Kohm

Better Trout Habitat: A Guide to Stream Restoration and Management
By Christopher J. Hunter

Beyond 40 Percent: Record-Setting Recycling and Composting Programs
The Institute for Local Self-Reliance

Coastal Alert: Ecosystems, Energy, and Offshore Oil Drilling
By Dwight Holing

The Complete Guide to Environmental Careers
The CEIP Fund

Death in the Marsh
By Tom Harris

Farming in Nature's Image
By Judith Soule and Jon Piper

The Global Citizen
By Donella Meadows

Healthy Homes, Healthy Kids
By Joyce Schoemaker and Charity Vitale

Holistic Resource Management
By Allan Savory

Inside the Environmental Movement: Meeting the Leadership Challenge
By Donald Snow

Last Animals at the Zoo: How Mass Extinction Can Be Stopped
By Colin Tudge

Learning to Listen to the Land
Edited by Bill Willers

Lessons from Nature: Learning to Live Sustainably on the Earth
By Daniel D. Chiras

The Living Ocean: Understanding and Protecting Marine Biodiversity
By Boyce Thorne-Miller and John G. Catena

For a complete catalog of Island Press publications, please write:
Island Press, Box 7, Covelo, CA 95428, or call: 1-800-828-1302